Taking the Fear Out of Data Analysis

The loan period may be shortened if the item is requested.

Taking the Fear Out of Data Analysis
A Step-by-Step Approach

Adamantios Diamantopoulos
and
Bodo B Schlegelmilch

Business Press
Thomson Learning™

Australia • Canada • Denmark • Japan • Mexico • New Zealand • Philippines

Puerto Rico • Singapore • South Africa • Spain • United Kingdom • United States

Taking the Fear Out of Data Analysis

Copyright © 2000 Thomson Learning

Business Press is a division of Thomson Learning. The Thomson Learning logo is a registered trademark used herein under licence.

For more information, contact Business Press, Berkshire House, 168-173 High Holborn, London, WC1V 7AA or visit us on the World Wide Web at:
http://www.thomsonlearning.co.uk

British Library Cataloguing-in-Publication Data
A catalogue record for this book is available from the British Library

---

ISBN 1-86152-430-7   1 0 0204 0479

First edition published 1997 by Harcourt Brace & Company Limited
Reprinted by Thomson Learning 2000

Typeset by Poole Typesetting Ltd., Bournemouth
Printed in Singapore by Kin Keong

Seriousness is the only refuge of the shallow.

Oscar Wilde

To the reader

The statistician is often thought as a man of another world as indeed are all mathematicians

W. J. Reichmann
Use and Abuse of Statistics
Pelican, 1961

Effective data analysis requires the effective use of statistics. Unfortunately, statistics is boring. It is boring to learn, it is boring to teach, and it is usually boring people who actually *like* statistics. Indeed, the comment that a statistician is 'a person who didn't have enough charisma to be a cost accountant', says it all!

Statistics is also hard. It is hard to learn, it is hard to teach (properly), and it is even harder to remember what little you may have learned. In short, statistics is rarely fun. But it *can* be — as you will soon find out. Trust us.

To our parents for believing in us,
to Heidi, Irene and Roger for living with us (*not* easy)
and to Jennifer for putting up with us

Instead of a preface

The preface world is an idyll peopled by super-intelligent conscientious colleagues, sweet-natured understanding families, and hyper-efficient telepathic typists. They all have a sense of humour which is 'unfailing' in the case of secretaries, and only the author ever makes mistakes.

Neil M. Kay
The Emergent Firm
Macmillan, 1984

The idea for this book was conceived a few years ago when Jennifer Pegg, our commissioning editor, approached us and asked us to consider writing a text on data analysis. (This was probably the biggest single mistake in her career, but never mind.) At the time we were both working at the University of Wales, Swansea and thought, 'Sure, why not?' However, upon giving the matter further thought, we decided that yet another conventional textbook on statistical analysis was not really needed (plenty of them are about). What *was* needed, we felt, was a structured guide that would exorcise the fear that many people have about numbers and statistics and would enable *anybody* to learn how to do data analysis. Such a guide should focus on understanding, rather than technical details, and be rather light-hearted as data analysis is not exactly a bunch of laughs. Much to our amazement, Jennifer agreed to go ahead with a book along these lines, and *Taking the Fear Out of Data Analysis* was born.

Then the problems started. First, upon seeing a couple of draft chapters, referees took exception to several of our jokes, found some of our examples politically incorrect, and queried the wisdom of not being sufficiently serious' (having said that, one referee admitted that he/she might be getting to be an 'old sourpuss'!). Next, one of the authors decided to take up a position in the United States, which didn't exactly help (all the faxes and e-mails in the world cannot substitute for being next door to each other). Finally, to top it all off, about three-quarters of the way through the book, the other author got an inflamed tendon from too much typing (yes, we did all the typing ourselves). If it wasn't for Jennifer's magic mix of constant encouragement with occasional threats of physical violence, we doubt whether this book would have been finished by now.

But it is not only Jennifer whom we want to thank. The anonymous reviewers (the identity of which Jennifer still refuses to reveal despite repeated attempts to bribe her) made many useful suggestions for improvement, all of which have been incorporated in the final draft. Several of our students in both the UK and the US kindly read chapters of the book and provided us with useful feedback; astonishingly, they did so voluntarily! Our partners, Heidi Winklhofer and Irene Schlegelmilch also found the time to read draft chapters, despite being very busy with their own careers. Johann Du Preez of the Department of Statistics at the University of Wales, Swansea was always willing to answer queries and clarify subtle technical points (see, there are some nice statisticians!). Gillian Hanna did more than her share of diagrams and photocopying of chapters. And, judging from their purrs while sleeping on it, the two cats Holly and Winkie also enjoyed the manuscript! Thanks to you all.

Adamantios Diamantopoulos
Bodo B. Schlegelmilch

Contents

To the reader — vii
Instead of a preface — ix
About the authors — xvii
Introduction — xxi

PART I: UNDERSTANDING DATA

1 What is *data* (and should you lose any sleep over it)? — 1

The nature of data — 1
Types of data — 4
Data and information — 7
Summary — 8
Questions and problems — 8
Notes — 9
Further reading — 9

2 Does *sampling* have a purpose other than providing employment for statisticians? — 10

The nature of sampling — 10
Sample selection — 12
Sample size determination — 16
The sampling process — 18
Summary — 19
Questions and problems — 20
Notes — 20
Further reading — 20

3 Why should you be concerned about different types of *measurement* and are some better than others? — 21

The nature of measurement — 21
Measurement scales — 23
Scaling formats — 30
Measurement error — 32
Summary — 37
Questions and problems — 37
Notes — 38
Further reading — 38

PART II: PREPARING DATA FOR ANALYSIS

4   Have you cleaned your data and found the *mistakes* you made?   39

The role of data cleaning   39
The role of data coding   43
Finding your mistakes   48
Transforming variables   49
Summary   50
Questions and problems   51
Notes   51
Further reading   52

5   Do you have access to a *computer package* or must you employ
Bill Gates?   53

The role of computers in data analysis   53
Data input packages   54
Data analysis packages   56
Presentation packages   58
Summary   60
Questions and problems   60
Further reading   61

6   Why do you need to know your *objective* before you fail
to achieve it?   62

The need for analysis objectives   62
Setting analysis objectives   63
The question of focus   64
Choosing the method of analysis   66
Summary   71
Questions and problems   71
Notes   72
Further reading   72

PART III: CARRYING OUT THE ANALYSIS

7   Why not take it easy initially and *describe* your data?   73

Purposes of data description   73
Frequency distributions   74
Grouped frequency distributions   77
Graphical representation of frequency distributions   81
Summary   88
Questions and problems   88
Notes   89
Further reading   89

8   Can you use few numbers in place of many to *summarize*
     your data?   90

     Characterizing frequency distributions   90
     Measuring central location   93
         The mode   93
         The median   95
         The mean   97
     Measuring variability   100
         The index of diversity   101
         The range and interquartile range   101
         The variance and standard deviation   102
     Measuring skewness and kurtosis   105
     Chebychev's theorem and the normal distribution   106
     Summary   114
     Questions and problems   115
     Notes   115
     Further reading   115

9   What about using *estimation* to see what the population
     looks like?   116

     The nature of estimation   116
     Setting confidence intervals   119
     Estimating the population proportion   121
     Estimating the population mean   123
     Estimating other population parameters   126
     Summary   127
     Questions and problems   128
     Notes   128
     Further reading   128

10   How about sitting back and *hypothesizing?*   130

     The nature and role of hypotheses   130
     A general approach to hypothesis-testing   135
         Step 1: Formulation of null and alternative hypotheses   136
         Step 2: Specification of significance level   137
         Step 3: Selection of an appropriate statistical test   140
         Step 4: Identification of the probability distribution of the
                 test statistic and definition of the region of rejection   143
         Step 5: Computation of the test statistic and rejection
                 or non-rejection of the null hypothesis   146
     Hypothesis-testing and confidence intervals   147
     Statistical and substantive significance   148
     Summary   150
     Questions and problems   150
     Notes   151
     Further reading   151

11	Simple things first: *one* variable, *one* sample	152
	Single sample hypotheses	152
	Assessing fit	153
	The one-sample chi-square test	154
	The one-sample Kolmogorov–Smirnov (K-S) test	157
	Testing for location	159
	The one-sample sign test	159
	The one-sample *t*-test	161
	Testing for variability	164
	Testing for proportions	165
	Testing for randomness	167
	Summary	169
	Questions and problems	170
	Notes	171
	Further reading	171
12	**Getting experienced: making *comparisons***	**172**
	The pleasure of comparing	172
	Independent measures: comparing groups	173
	The two-sample chi-square test	175
	The *k*-sample chi-square test	179
	The Mann–Whitney *U* test	180
	The Kruskal–Wallis (K-W) one-way analysis of variance (ANOVA)	182
	The two-sample *t*-test	184
	One-way analysis of variance (ANOVA)	187
	Related measures: comparing variables	190
	The McNemar test	192
	The Cochran *Q* test	193
	The paired-sample sign test	193
	The Friedman two-way analysis of variance (ANOVA)	194
	The paired-sample *t*-test	195
	The t_r-test for all pairs	196
	Summary	196
	Questions and problems	196
	Notes	197
	Further reading	197
13	**Getting adventurous: searching for *relationships***	**198**
	The mystique of relationships	198
	Measures of association	199
	Cramer's *V*	199
	Spearman's rank-order correlation	201
	Pearson's product moment correlation	203
	Correlation and causality	206
	Summary	207
	Questions and problems	207
	Notes	208
	Further reading	208

14 Getting hooked: a look into a *multivariate* future   209

    The nature of multivariate analysis   209
    Types of multivariate techniques   211
    Dependence methods   214
        Multiple regression analysis   214
        Multiple discriminant analysis   214
        Canonical correlation analysis   214
        Multivariate analysis of variance   215
    Interdependence methods   216
        Factor analysis   216
        Cluster analysis   217
    Summary   217
    Questions and problems   217
    Further reading   218

15 It's all over . . . or is it?   219

    The written research report   219
    The oral presentation   221
    Summary   224
    Questions and problems   225
    Further reading   225

Index   227

About the Authors

Adamantios Diamantopoulos BA, MSc, PhD, MCIM is Professor of Marketing and Business Research at Loughborough University Business School and also Research Director of the School. He was previously Professor of International Marketing at the European Business Management School, University of Wales, Swansea where he headed the Marketing Group. Other past academic posts include full-time appointments at the Universities of Edinburgh and Strathclyde and visiting Professorships at the Universities of Miami, Strasbourg and the Vienna University of Economics and Business Administration. His research interests are in pricing, sales forecasting, marketing research and international marketing. He is the author of numerous academic publications and is a member of the editorial review boards of several prestigious academic marketing journals. He is also Associate Editor of the *International Journal of Research in Marketing*. When not working, he spends most of his time with his horse, Legaun's Legend. He also enjoys riding motorcycles, skiing and having enormous Chinese meals. His mother is worried that he will never grow up.

Bodo B Schlegelmilch Dipl. Betriebswirt, MSc, PhD, FCIM, MMRS is Professor of International Marketing Management at the Vienna University of Economics and Business Administration. He was previously Professor of Marketing at Thunderbird, The American Graduate School of International Management in Arizona, where he headed the Marketing Group and founded the US Government supported Institute for International Business Ethics. He has also held posts as British Rail Chair of Marketing at the University of Wales, Swansea and as lecturer of Marketing and International Business at the University of Edinburgh. Other academic and commercial posts include the University of California at Berkeley, the University of Miami, Deutsche Bank and Procter & Gamble.

Professor Schlegelmilch is President of Canyon Consulting Inc. and has undertaken consulting and executive education work in strategic marketing for major multinationals, including Allied Signals, Anheuser Busch, AT&T, BellSouth, Black & Decker, Cable & Wireless, EDS, Estee Lauder, Goldman Sachs, MCI, Merck Sharp & Dohme, KPMG, Philip Morris, Schlumberger, Sunkyong, Universal Flavors and Pharmacia & Upjohn. He has also taught in international marketing programmes in Austria, Britain, Egypt, France, Germany, New Zealand and Russia, has published over 100 academic papers on international marketing and management issues, and is on the editorial boards of seven leading academic journals.

When not researching, writing or teaching, Professor Schlegelmilch enjoys mountain-biking with his son, hanging out at nice beaches, windsurfing, eating Japanese meals and writing long and complicated autobiographical statements. His wife and mother both know he will never grow up.

From the Pre-publication Reviews

"While most people make decisions daily without thinking too deeply about the process, present them with a collection of quantitative data to analyse and their eyes glaze over from lack of comprehension and fear. *Taking the Fear Out of Data Analysis* addresses this issue head on and, in my view succeeds brilliantly.

In a clear and amusing style the authors lead the reader through a step-by-step explanation of the principles of data analysis. For the novice about to embark on formal data analysis for the first time this book explains what needs to be done, why and, most importantly, how.

For the more experienced researcher in need of accessible advice on procedures and techniques each chapter provides a synoptic overview in every day language that avoids the obfuscation (intentional or otherwise) that makes traditional texts so difficult to understand.

This book is essential reading for any student or practitioner faced with the task of completing a project involving the collection, analysis, and presentation of data."

Professor Michael J Baker, University of Strathclyde

"A very entertaining and easy to read book on the often difficult and obtuse topic of data analysis. An excellent resource for the analytical neophyte and for those who have always tried to avoid statistical literacy because of their fear of formulas. The authors elegantly, yet humorously achieve their goal of 'taking the fear out of data analysis'."

Professor Kenneth R Ferris, Thunderbird –
The American Graduate School of International Management

"This book is a real gem. It takes a notoriously difficult subject (both to understand and to apply) and gently leads the reader by the hand from the very basics to an advanced level. Peppered throughout with humorous examples, the authors have written a text that will find a ready market on both degree and graduate courses, especially amongst those with a pathological fear of numbers but who know that they need to understand the basics.

Despite its often witty and sometimes whimsical approach this is a solid text, written by two of the best data analysts in the business. They certainly know their stuff and even seasoned researchers will find the 'warnings' and 'hints', tables and figures, timely reminders of things they may have forgotten (or perhaps, perish the thought, never even knew).

Taking the Fear Out of Data Analysis does just that. Now no-one has any excuses for fainting at the mention of statistics!"

Professor Graham Hooley, Aston Business School

"The authors have tackled a topic of critical importance. The text is easy and pleasant to read, following a challenging format, lively and concise, written in a light, refreshing and amusing manner, yet introducing a very robust content. The analytical discussion of the techniques is comprehensive, incorporating many good examples, hints and warnings. The book provides a smooth 'route planner' to data analysis and really takes the fear out of it. Moreover, it is painless! . . ."

Professor Luiz Moutinho, University of Glasgow Business School

"EXCELLENT! The authors are to be commended"

Peter Mudie, Napier Business School

"This is a very human approach to the inhuman task of handling statistics. It should be read by management students and managers before and after handling data, and should dramatically clarify the data analysis task for them."

Nigel F Piercy, Sir Julian Hodge Chair of Marketing and Strategy,
Cardiff Business School, University of Wales

"*Taking the Fear Out of Data Analysis* is more than just another book on data analysis. It is actually fun to read. The approach is thoughtful, intelligent, practical, thorough, helpful and even humorous. The humour helps break down barriers and keeps things refreshing. It reduces tension and even fears.

It is obvious that the authors know what they are talking about – the book is loaded with examples and very helpful hints and warnings. The book is also organized in a logical way which makes it easy to move from chapter to chapter. This book makes data analysis concepts and techniques easier to learn (or review)."

Professor David Ricks, Thunderbird –
The American Graduate School of International Management

"*Taking the Fear Out of Data Analysis* will have a huge influence on the way we do research in Britain. It is a book we need. It will not only find its way on to the bookshelf of any serious social science or business researcher, but it is one that will be taken down and consulted regularly. The only danger is that too many people will want to borrow it. I would recommend binding it in plain brown paper.

Besides being the duo with the longest names around, Bodo Schlegelmilch and Adamantios Diamantopoulos are leaders in rigorous methodology in social science research. They are lovers of research and the joy they find in it is clearly demonstrated in the book. It is a book to read as well as to consult, to enjoy as well as to learn from."

Professor John Saunders, Loughborough University Business School

". . . hidden behind some bizarrely memorable examples and illustrations is a very fine introduction to data analysis. This book will be of value to those approaching quantitative analysis for the first time, and should be on the reading list for project-based courses and for new research students."

Richard Speed, The University of Melbourne

Introduction

The trouble with numbers is that they frighten a lot of people

Leslie W. Rodger,
Statistics for Marketing
McGraw-Hill, 1984

This book has been written for people who do not *like* data analysis, who do not *want* to become analysts or statisticians, but who *have* to learn about data analysis for whatever reason (e.g. to pass a course at college, complete a dissertation, or get/keep a job). It has also been written for people who *think* they cannot understand data analysis and statistics, having had bad experiences with textbooks full of formulae and little substance, and teachers full of confidence and no humour. In fact, this book has been written for anybody suffering from the 'I hate numbers' syndrome — and we know there are plenty of you out there.

What we tried to do in this book is quite simple: take your hand and lead you through the entire data analysis process without boring you to death along the way. Our specific aims have been threefold:

1. To provide a comprehensive but digestible *introduction* into the strange world of data analysis, assuming no prior knowledge on your part.
2. To indicate the *linkages* among the various stages of the data analysis process and highlight the implications of good/bad early decisions on subsequent ones.
3. To demonstrate that learning about data analysis can be an *enjoyable* experience; hopefully, after you have finished reading this book, you will feel that way too.

Our philosophy behind the content and structure of the book is based on a few basic premises. First and foremost, our main concern is with *understanding* rather than memorizing. Thus we urge you to channel your learning effort towards grasping and digesting the various concepts/techniques of data analysis rather than mechanically reproducing a bunch of formulae. Moreover, at this introductory level, we feel that your attention should be directed towards *key issues* and *major building blocks* rather than statistical refinements and details. Consistent with this view, we keep the number of formulae down to an absolute minimum and do not bother with providing mathematical proofs (the latter being a sure way of sending you to sleep!). We also firmly believe that a point is best driven home by means of an *example*. Consequently, we make liberal use of (mostly silly) examples and illustrations to show how the various concepts and techniques can be applied. Lastly, we see no harm in making you smile from time to time — a hefty dosage of humour is often the only way to keep one sane while learning/doing data analysis!

You can use this book in a number of ways, depending upon your background and objectives. For those of you with no prior experience of data analysis and little, if any, statistical knowledge, we strongly recommend that you go through the book chapter by chapter. In this way you will be introduced to more complex material in a gradual manner and should not feel lost at any time. On the other hand, those of you with some previous exposure to research methods and/or statistics may opt for a more flexible approach; for example, you may wish to concentrate on Chapter 4 (dealing with data preparation and coding) onwards, and refer back to Chapters 1–3 (which are really background chapters) on an 'as required' basis. Finally, for those of you who are particularly interested in specific types of analysis, you should primarily focus on Chapters 7–14, which cover different analytical techniques (and presuppose familiarity with basic data analysis principles).

There are also some key chapters that everyone should read. These include Chapter 6 (on setting analysis objectives), Chapter 9 (on the nature of statistical estimation), Chapter 10 (on the principles of hypothesis-testing) and Chapter 15 (on evaluating and presenting the analysis). Moreover, all readers would do well to heed the numerous **HINTS** and **WARNINGS** dispersed throughout the various chapters. These serve to emphasize key points, awareness of which can prevent problems and make life easier for you.

The Further Reading section at the end of each chapter should also be consulted. Rather than provide a long list of references, we have intentionally limited ourselves to a selection of a few key sources which we feel best amplify and complement the material covered in the chapter. Each suggested reading has been briefly annotated and there is no duplication of sources in the Further Reading sections of the various chapters. Having said that, many of the suggested readings may also be useful for issues discussed in the other chapters — so, keep an open mind and be flexible.

The book is organized into three main parts, containing a total of 15 chapters.

Part I, *Understanding Data*, provides the necessary background by looking at the nature of data, the sampling process and the notion of measurement. These are essential building blocks underpinning data analysis and a prerequisite for understanding the application of statistical techniques.

Part II, *Preparing Data for Analysis*, focuses on the various tasks associated with converting raw data into a form that can be analysed and on setting objectives for the analysis. Careful attention to the issues raised here will prevent a lot of problems at the actual analysis stage.

Part III, *Carrying out the Analysis*, examines the rationale behind different types of analysis and introduces a wide variety of analytical methods appropriate for different circumstances. Starting with simple approaches to describing and summarizing data, a number of techniques are considered which, if properly applied, will enable you to get the most out of your data. A first taste of multivariate analysis is also provided, although at a very basic level. Finally, several issues are raised relating to the evaluation and presentation of a data analysis project and the preparation of written and oral research reports.

By the time you finish reading this book you should be in a position to know *what analysis* to apply, for *which purpose* to *what kind of data*. The chapter-by-chapter overview that follows should help clarify what all this means.

Chapter 1 lays the groundwork for the rest of the book by introducing you to the **data matrix**, which is the raw material you will work with whenever you do data analysis. Here you will get acquainted with **units of analysis**, **variables** and **values**, which are the essential ingredients of data. You will also learn about different **types of data** and about the distinction between data and **information**. By the time you finish this chapter, you should be able to talk about data as if you knew something about it!

Chapter 2 looks at **sampling** to demonstrate how units of analysis may come about in a particular project. You will understand the rationale for taking a **sample** rather than studying the whole **population** and the different **sampling methods** that you can use. You will also encounter the concept of **sampling error** which, no doubt, will haunt you forever after! Finally, the numerous considerations for determining **sample size** will be piled upon you — if only to confuse you even further!

Chapter 3 describes what **measurement** is, why it is important and how it can be done. You will recognize the advantages and disadvantages of different **measurement scales** and get to know a variety of **scaling formats.** Following this, you will encounter a second kind of error, the notorious **measurement error**, which will also haunt you forever after (either with the sampling error or on its own). By the end of this chapter, you should have a firm grasp of the options available for measuring your variables and the interpretation of the resulting values.

Chapter 4 moves into the practicalities of preparing data for analysis, focusing on the process of **data cleaning**. You will learn how to detect the many errors you have made by using different **edits** and how to deal with ambiguous, inconsistent, and missing data. You will also be shown how **data coding** can best be accomplished and how to draw up a **code book** prior to inputting data into your computer. Lastly, you will see how easy it is to augment your data set by creating new variables through intelligent **variable transformations**.

Chapter 5 praises the miracles of modern technology — **computer packages** — which do most of the hard work for you. Starting with **data input packages**, you will be introduced to different options for getting your coded data into a computer. Next, you will get an overview of the various types of **data analysis packages** which you can use to crunch your data (once you have mastered the principles of data analysis, of course). Lastly, you will be shown how to impress your audience endlessly by using flashy **presentation packages**. If nothing else, by becoming familiar with computer packages, you should be in a much better position to blame them if things go wrong!

Chapter 6 emphasizes the need for having clear **analysis objectives** before embarking on the actual analysis (otherwise you will not know how to start or when to stop). These will ensure that your analysis is relevant, comprehensive and efficient. You will be introduced to different analytical perspectives, namely **description**, **estimation** and **hypothesis-testing** (which will be dealt with in more detail in later chapters). The chapter will conclude with a discussion of the factors governing the choice of the **method of analysis** (which will undoubtedly become the subject of your worst nightmares for years to come!).

Chapter 7 focuses on **data description**, which is usually the first type of analysis that you will want to do. You will get to know the various forms of **frequency distributions** and the steps to take in grouping data. Your artistic talents will also be thoroughly stimulated by an examination of different types of **graphical displays**. At this stage you will wonder what the purpose of life would be without percentiles, true class limits and ogives!

Chapter 8 soldiers on with data description and introduces you to different **summary measures** which can be used to capture **typical** or **average** responses as well as the extent of **variability** in your data. By using different summary measures in conjunction with one another, you will be able to identify the **shape** of a frequency distribution and compare it to known forms. In this context, the famous **normal distribution** will serve as a useful point of reference.

Chapter 9 deals with the process of **estimation** and shows how you can talk about the population when you only have a sample. First, you will become familiar with the concept of a **sampling distribution** and then learn how to set **confidence intervals** for different **population parameters**. Being able to make inferences from a sample to a population will surely set you apart from the uninitiated punter!

Chapter 10 ventures into the mystical world of **hypothesis-testing** and provides you with an understanding of the basic principles associated with developing and testing specific research propositions. You will learn about different **types of hypotheses** and the rationale behind **significance testing** which, like estimation, also enables you to make inferences from a sample to a population. Concepts like 'null hypothesis', 'p-values' and 'regions of rejection' will become second nature to you and a topic of conversation at every possible opportunity!

Chapter 11 deals with the simplest type of hypotheses, namely those involving a single variable and a single sample. Here you will be shown how to examine the **fit** of your frequency distributions against prior expectations or against a theoretical distribution. In addition, you will be able to determine whether your sample is likely to have been drawn from a population with known **parameter values** and whether your sample is in fact **random**. By the time you complete the chapter, you should feel confident enough to face more complex hypotheses (involving more than one sample or more than one variable).

Chapter 12 extends your journey into hypothesis-testing by concentrating on **comparisons**. First, you will learn how to compare two or more groups on the same variable — what is known as an **independent measures** (or samples) comparison. Next, you will address the issue of comparing two sets of responses from the same group, involving a **related measures** (or samples) comparison. In both cases, you will be testing to see whether **significant differences** exist, that is whether any observed differences on your sample results are likely to reflect 'true' differences in the population. By becoming an expert on comparisons, you will be able to impress your friends with profound statements of the sort: 'Basketball players are, on average, significantly taller than jockeys' and 'There is no significant difference between the proportions of male and female construction workers with a passion for ornithology'.

Chapter 13 rounds up the analysis by showing you how to investigate **relationships** between two variables. Here you will be exposed to **measures of association** which can tell you whether two variables are related to one another and which can be subjected to significance tests to see whether any observed relationship (based on your sample data) is also likely to hold in the population. Once you have established a **significant relationship**, your association measure will also enable you to assess its **strength** and **directionality** (i.e. positive or negative).

In Chapter 14 we show you that there is *much* more to data analysis than is covered in this book — **multivariate analysis** procedures open up a whole lot of new opportunities for extracting information from your data. Here we only give you a brief description of what different multivariate techniques can do for you and completely ignore the technical issues involved. Advanced data analysis using multivariate procedures cannot be done justice in a single chapter — maybe we'll write a book about it one day!

Chapter 15 is an oasis of sanity at the end of your data analysis odyssey. Unbelievable as it may sound, in this chapter you will *not* have to grasp new theoretical concepts, learn yet another technique, or interpret more statistical results. Responding to your pleas for mercy, the purpose of this final chapter is to make you sit back and think about the 'Now what?' question. Here we talk you through how to **present** your analysis to your audience(s): Is it too technical? Are the practical implications clear? Is the presentation attractive? Unless the presentation of your analysis (written and/or oral) is effective for the *specific* audience involved, all your efforts will have been in vain. Not only must you do the right thing and do it right: you must convince *others* that you have done so.

Have fun.

PART I:

UNDERSTANDING DATA

What is *data* (and should you lose any sleep over it)?

The Nature of Data

Few of us would disagree that an inescapable (and extremely boring) part of modern life is form-filling. We have to fill in forms for all sorts of things, for example to apply for a job or a place at college, to obtain a driving licence or a passport, or to be given a credit card or loan. In doing so, we routinely supply details about ourselves such as age, sex, family size, residence and income, to name but a few.

As if form-filling was not enough, we are also often accosted in the street by extremely 'friendly' (and sometimes obnoxious) characters and are asked to 'please answer a few questions' regarding our shopping habits (e.g. How often do you shop at *El-Cheapo* supermarket?), product preferences (e.g. Do you prefer *Superclean* over *Ultrasteril* washing powder?), voting intentions (e.g. If a General Election were called tomorrow, would you vote for (a) *The Conversation Party*, (b) *The Favour Party*, or (c) *The Literal-Demagogue Party*?), or opinions regarding the Single European Market (e.g. Do you feel that British producers of eucalyptus-flavoured dog biscuits have benefited from EU membership?). To top it all off, even the privacy of our own home is not enough to prevent the hungry information-seekers from reaching us. How many times did we have to miss a crucial part of *Neighbours* or *Eastenders* in order to answer the phone, only to find out that the caller is 'Belinda Pain, from *Persistence Research, Inc.*, and was wondering whether you would participate in a survey regarding time-share holidays in Tirana'?

All the above situations have one thing in common: they are attempts to collect **data**. While the *approach* used to obtain the data varies in each instance, the *objective* is always to secure responses from an individual with regard to certain characteristics of interest. Now in statistical jargon (yes, you *do* have to learn some of it), the individuals approached are called **units of analysis** (or sometimes 'observations', 'cases' or 'subjects'), the characteristics studied are termed **variables** and the responses linking the individuals to the characteristics are known as **values**. Together, units of analysis, variables and values make up what we call 'data'. Thus when we refer to data, we implicitly address three distinct issues, notably (a) the *topic* of interest (as described by the set of variables), (b) the *respondents* (as indicated by the units of analysis), and (c) the *responses* of the latter in relation to the topic of interest (as reflected in the values of the variables).

data

units of analysis
variables
values

constant

Variables can assume different values for different units of analysis; if this is not the case (i.e. when all units of analysis have the same value), then we are not dealing with a variable but with (surprise, surprise) a **constant**.

In principle, units of analysis do not *have* to be individuals (although in a great deal of social research they are). They can be objects (e.g. beer brands, car types, computer makes), time periods (e.g. months, years, decades), events (e.g. strikes, accidents, shareholder meetings), or other entities (e.g. firms, cities, nations). Neither do variables have to refer to human properties; they can be product features (e.g. speed, durability, colour), organizational dimensions (e.g. centralization, formalization, span of control), national characteristics (e.g. inflation rates, government spending, interest rates), in fact *anything* that can be used to characterize the particular unit of analysis.

data set

When we have a number of units of analysis (e.g. 200 first-year parapsychology students), a number of variables (e.g. age, sex, parents' income, nationality) and a set of values linking the units of analysis to the variables, the result is a **data set**. This can be best visualized as a matrix, the rows of which represent the units of analysis, the columns the variables and the matrix cells the relevant values; for our example, the idea is shown in Table 1.1.

Table 1.1 An example of a data matrix

Units of	Variables			
analysis	Age	Sex	Parents' income	Nationality
Student 1	17 years	Male	£36 000	British
Student 2	18 years	Male	£22 000	Malaysian
Student 3	20 years	Female	£85 500	Costa Rican
•	•	•	•	•
•	•	•	•	•
•	•	•	•	•
Student 200	19 years	Female	£17 000	British

data matrix

If there are n units of analysis (respondents, objects, events, etc.) and m variables, the **data matrix** looks like Table 1.2, where R_{ij} is the response that unit i gives to variable j (in other words, R_{ij} is the value for unit i on variable j). The subscripts i and j are simply used for counting purposes; in other words, $i = 1,2, \ldots ,n$ and $j = 1,2, \ldots ,m$. Obviously, depending upon how many units of analysis and variables we have, n and m will vary; for example, in Table 1.1 $n = 200$ and $m = 4$.

The main benefit from arranging data in matrix form is that the three-fold nature of data (i.e. units of analysis, variables and values) becomes immediately visible; in fact, *all* data sets can be represented in the form of Table 1.2.

The data matrix is the starting point for analysis and its structure determines the kind of analysis that can be legitimately carried out. Specifically, the number of rows, n, indicates how many units are being studied, i.e. the

sample, population

sample or **population** size (we will have much more to say about samples and populations in Chapter 2). The number of columns, m, on the other

Table 1.2 General form of a data matrix

Units of analysis	Variables											
	V_1	V_2	V_3	•	•	•	•	V_i	•	•	•	•	V_m
O_1	R_{11}	R_{12}	R_{13}	•	•	•	•	R_{1i}	•	•	•	•	R_{1m}
O_2	R_{21}	R_{22}	R_{23}	•	•	•	•	R_{2i}	•	•	•	•	R_{2m}
O_3	R_{31}	R_{32}	R_{33}	•	•	•	•	R_{3i}	•	•	•	•	R_{3m}
•	•	•	•					•				•
•	•	•	•					•				•
•	•	•	•					•				•
O_i	R_{i1}	R_{i2}	R_{i3}	•	•	•	•	R_{ij}	•	•	•	•	R_{im}
•	•	•	•					•				•
•	•	•	•					•				•
O_n	R_{n1}	R_{n2}	R_{n3}	•	•	•	•	R_{nj}	•	•	•	•	R_{nm}

hand, indicates how many variables are used to characterize the units of analysis; depending upon the number of variables, k ($k \leq m$), that are *simultaneously* manipulated by applying statistical techniques, we talk about **univariate** ($k = 1$), **bivariate** ($k = 2$), and **multivariate** ($k > 2$) data analysis, respectively (we shall return to this issue in Chapter 6). Finally, the natures of the values, R_{ij}, reflect the **level of measurement** of the variables and thus indicate what can and what cannot be said about the units of analysis (measurement will be dealt with in some detail in Chapter 3, so don't worry if you feel totally confused at the moment!).

> **univariate data analysis,**
> **bivariate data analysis**
> **multivariate data analysis**
> **level of measurement**

Now, here's something you must always watch out for when you are dealing with a data set which is based on questioning respondents: *a question and a variable are not necessarily the same thing.* Very often answering what appears to be a single question in a questionnaire may in fact require **multiple responses** and will thus result in a number of variables. Table 1.3 illustrates this point: irrespective of whether Question A or Question B is asked, the response alternatives are identical. However, the nature and number of the resulting variables are not. If Question A is asked, then only *one* option needs to be ticked to answer it. Consequently, to capture all possible responses, a single variable would be sufficient; this would be called something like 'most preferred publication' and would take five values, one for each publication involved (e.g. 1 = *Wall Street Journal*, 2 = *The Times*, . . . ,5 = *Unspeakable Acts*). If Question B is asked instead, a single

> **multiple responses**

Table 1.3 Single- and multi-response questions

A. If you only had enough money to subscribe to just one of the following publications, which one would you choose?

B. If money were not a problem, which of the following publications would you subscribe to?

	(Please tick)
Wall Street Journal	[ ]
The Times	[ ]
Journal of the Mathematically Insane	[ ]
Mongolian Economic Review	[ ]
Unspeakable Acts	[ ]

variable is *not* sufficient to capture all possible responses, because a respondent may wish to subscribe to more than one publication (e.g. *The Wall Street Journal* and *Unspeakable Acts*) and thus legitimately tick multiple options. In this case, in order to ensure that all possible response are captured, five variables would be needed (i.e. one for each publication). The reason for this is that Question B is not really a single question but a *series* of questions of the form: 'Would you subscribe to the *Wall Street Journal?*', 'Would you subscribe to *The Times?*', etc. The response to each of these questions would be of the 'yes/no' variety, resulting in five variables, each taking two possible values (e.g. 1 = would subscribe, 2 = would not subscribe).

This brings us to the first WARNING and first HINT of this book (many more will follow, don't you worry!).

Types of Data

Let us now move on to the different *types* of data that one may come across in a data analysis project. While there are many data classification schemes, at this stage we shall briefly look at different types of data according to (a) their meaning, (b) their source, and (c) their time dimension; in Chapter 3 a fourth classification of data will be introduced based upon their measurement properties.

facts   Focusing initially on different kinds of data according to their meaning, there are data that refer to **facts**, that is, characteristics or situations that exist or have existed in the past. Things like age, sex, income, church membership, 1973 sales of Wartburg automobiles, and number of drunken Taiwanese visitors at last year's *Oktoberfest* in Munich, are all examples of facts. Descriptions of individuals' present behaviour (e.g. current shopping habits) and past behaviour (e.g. historical voting patterns) also fall in this category.

awareness, knowledge   Secondly, there are data that refer to **awareness** or **knowledge** of some object or phenomenon. Typical examples here are brand awareness (e.g. Which of the following toothpaste brands do you recognize: (a) *Draculadent*, (b) *Vampirmed*, (c) *Ghoulshine?*); knowledge of important events (e.g. when did Princess Anne first wear a purple mini-skirt in public?); and mastery of a certain subject or topic (e.g. Who wrote *Das Kapital*: (a) Woody Allen, (b) Jeffrey Archer, (c) Stephen King, (d) Karl Marx, or (e) Karl Marx and Woody Allen together?).

intentions   Thirdly, there are data representing **intentions**, that is acts that people have in mind to do (i.e. their anticipated or planned behaviour). Such intentions can relate to future purchasing behaviour (e.g. Having tried your free sample of *Explosion* laxative, do you intend to buy it in the future?); social behaviour (e.g. Mrs Corleone, you will be sorry to hear that your son has failed all his final exams. Do you intend to (a) let him get away with it, (b) cut his pocket-money by 96%, or (c) confiscate his Ferrari?); and personal behaviour (e.g. Mr Gorbachev, having just been made redundant, do you intend to (a) become a traffic warden, (b) live off your pension, or (c) sell everything you have and become a Moonie?).

attitudes, opinions   Fourthly, there are **attitudes** and **opinions** data, which indicate the views, preferences, inclinations or feelings of people towards some object or phenomenon. Examples of attitude/opinions data are product /service evaluations (e.g. Do you think that Railtrack safety record is (a) brilliant, (b) good, (c) poor, (d) abysmal?); political beliefs (e.g. In your view, does the Labour or the Conservative Party have the better policy for providing employment opportunities for single teenage mothers in the Outer Hebrides?); and views on social issues (e.g. What is the single most important problem facing mankind today: (a) poverty, (b) drugs, (c) AIDS, (d) the Welsh rugby team's recent 'bad streak').

Lastly, there are data relating to the **motives** of individuals. Motives are internal forces (i.e. desires, wishes, needs, urges, impulses) that channel behaviour in a particular way. Although motivations may be complex and difficult to articulate, data of this kind are quite important because they can tell us *why* people behave in the way they do. Examples include reasons given for doing or not doing a certain thing (e.g. 'I go to aerobics so that my fantastic body becomes even more irresistible'; or, 'I don't do any exercise whatsoever because I'm a lazy slob'); explanations for preferring something over something else (e.g. 'I prefer Maggie Thatcher running the country rather than Labour because of her strong pro-European views'); and rationales for holding certain views or opinions (e.g. 'I think that memorizing statistical formulae is a total waste of time, because you can look them up in a book – or use a computer to do the work instead'). **motives**

Turning attention to types of data according to their source, a broad distinction can be drawn between **primary** and **secondary data**. Primary data are data collected with a specific purpose in mind, i.e. for the needs of a particular research project. Such data are usually gathered by the researcher via **surveys** (conducted face-to-face, by telephone, or through the mail), **experiments** (carried out in the laboratory or in a 'natural' setting), or **observation methods** (using mechanical devices or humans to record observed behaviour). In contrast, secondary data are data which have not been gathered expressly for the immediate study at hand but for some other purpose; such data, however, *might* be of relevance/use for a particular research project (in other words, somebody else has done the work but you may be lucky enough to be able to use it without getting your hands dirty!). A wealth of secondary data can be found in any halfway decent library in the form of **published statistics** (by government departments, trade associations, chambers of commerce, and research foundations), **annual reports** (published by business firms as well as non-profit organizations), and **abstracting & index services** (covering thousands of periodicals, academic journals and newspapers). For those who can afford to pay for them (and beware, because they don't come cheap!), there are also **syndicated services** (providing regular detailed information on a particular country/industry/product group) and **database services** (providing tailor-made mailing lists, allowing fast access to computerized information sources worldwide, or enabling 'electronic' transfer of data sets from one location to another). **primary data, secondary data**

surveys
experiments
observation methods

published statistics

annual reports
abstracting & index services

syndicated services
database services

The final classification of data to be considered has to do with the time dimension and distinguishes between data relating to a single point in time and data relating to a number of time periods. Data of the former type are known as **cross-sectional data** while the latter are commonly referred to as **longitudinal data**. From an analysis point of view, the distinction **cross-sectional data**
longitudinal data

		Data set	
Variables	A	B	C
1. February 1996 purchases of *Glennfiddle* Scotch	15		12
2. February 1996 purchases of *Johnny Stalker* Scotch	10		8
3. February 1996 purchases of *Castledrain XXXX* Lager	5		7
4. March 1996 purchases of *Glennfiddle* Scotch		18	14
5. March 1996 purchases of *Johnny Stalker* Scotch		11	10
6. March 1996 purchases of *Castledrain XXXX* Lager		8	6

All purchases are measured in 1000 litres!

between the two is quite important because it determines whether inferences regarding *change* can be made. To illustrate this, consider for a moment the three data sets displayed in Table 1.4; the units of analysis in all cases consist of 600 insomniacs living in Auchtermuchty, Scotland.

Data sets A and B are examples of cross-sectional data. Each provides a *snapshot* of the variables of interest at a particular point in time; in this instance, data set A informs us about whisky and lager purchases in February 1996 while data set B does the same thing for March 1996. *Within* each data set, one could compare purchases across product types and reach conclusions regarding the most popular drink in a particular time period.

Data set C, on the other hand, is an example of longitudinal data and involves *repeated* measurements over time on the variables of interest; data set C informs us about whisky and lager purchases in February *and* March 1996. As a result, one can not compare only purchases across product types but also purchases of the *same* product *over time*; thus, in addition to being able to draw the kind of conclusions that data sets A and B enable, conclusions regarding *changes* in the relative popularity of the three drinks are now possible.

Now, having just distinguished between cross-sectional and longitudinal data, we shall immediately confuse you by suggesting that one can do longitudinal analysis by using cross-sectional data! This is not as impossible as it first sounds. Take, for example, data sets A and B and *combine* them, i.e. imagine that they are parts of the same study. Piecing the two data sets together provides information on the same variables (here whisky and lager purchases) of different (but comparable) units of analysis (here *two* lots of 600 Auchtermuchty insomniacs) at different points in time (here February and March 1996 respectively). Data of this kind are known **trend data** as **trend data** and enable inferences to be drawn regarding changes in *aggregate* behaviour, attitudes, etc. Election polls provide a good illustration in this context: '. . . a poll commissioned by the *Daily Polygraph*, in which the voting intentions of a nationally representative sample of 11 893 birdwatchers were obtained yesterday, shows *Favour* standing at 43%, the *Conversation Party* at 37% and the *Literal-Demagogues* at 20%. A similar poll, conducted two weeks ago by the *Financial Crimes*, had *Favour* neck and neck with the *Conversationists* at 44.3% and 44.1%, respectively, with the *Literal-Demagogues* trailing at an appalling 11.6%'!

Trend data should be distinguished from true longitudinal data (such as those provided by data set C) where the *same* units of analysis are studied at different points in time. The latter are sometimes also referred to as **panel data** and, in addition to capturing aggregate changes over time, they enable inferences to be drawn as to changes in *individual* behaviour. A typical example of this kind of data is provided by consumer panels, in which a number of individuals or families (usually balanced on such variables as age, income and geography) record their purchases of a number of products at regular intervals (e.g. monthly). Their records are subsequently used, among other things, to determine the degree of brand loyalty (i.e. the proportion of those buying brand X in period 1 that also bought brand X in period 2) and degree of brand switching (i.e. the proportion of those buying brand X in period 1 that bought some other brand in period 2). As you have probably fallen asleep by now, here's a warning to wake you up.

In general, given a set of variables, four basic kinds of studies can be distinguished according to (a) whether the variables concerned are measured once or repeatedly, and (b) whether the same or different units are studied in each case (Table 1.5).

panel data

WARNING 1.2

Unless you are certain that the *same* units of analysis have been measured over time on the same variable, then conclusions regarding change at the *individual* level are *not* permissible.

Units studied	Points in time for observations	
	One	Many
Same	Cross-sectional study	Panel study
Different	Cross-sectional replication	Trend study

Table 1.5 Basic types of studies

As you might have suspected, there are more types of data sets than those we have discussed. The good news is that, for the most part, they represent different combinations or variations of the basic types shown in Table 1.5. For example, the well-known **experimental design** of the 'before–after' variety with a control group is essentially a combination of two panels, one of which has been unlucky enough to be exposed to the experimental treatment (e.g. subjects were forced to watch 43 TV adverts of *Loch-Ness-Café Gold Bland*) and one which has been spared the torture (i.e. subjects were left in peace). Similarly, an **omnibus panel** (no, this is not the rear section of a London double-decker bus), is essentially a series of cross-sectional studies, in which the same group of individuals (i.e. the panel members) is measured on different variables at different points in time (e.g. at one time, panel members may be asked to evaluate alternative packages for cat food and, at another time, to indicate their attitudes towards a new type of heavy-duty flea spray). Often, a sub-group of the total panel is selected for a particular purpose; for example, if one wanted to study consumer reactions to a new brand name for hair-removing wax, only those panel members that are female and over a certain age (e.g. sixteen plus) would normally be surveyed.

experimental design

omnibus panel

Data and Information

Before we move on to even more exciting stuff (!), we need to look briefly at the relationship between data and **information**. In everyday language,

information

the two are usually taken to be synonymous; however, one can distinguish between them in at least two senses.

Firstly, one can look at information as the *product* of data, i.e. information is data which has been digested and analysed. In other words, information is the knowledge obtained and conclusions arrived at after appropriate analytical techniques have been applied to the data matrix in Table 1.2. Arguably, a mass of **raw data** (i.e. unsummarized/unstructured) is of little informational value in itself. Imagine, for example, 6000 questionnaires on reading habits landing on your desk and, five minutes later, your (horrible) boss asking you: 'So Mr Bigbrains, where should we advertise our new bubblebath?'. Indeed, 'to be presented with a mass of unsummarized data is like being abandoned in the depths of a dense forest without the aid of a compass'.[1]

raw data

A second distinction between data and information can be drawn on the basis of *relevance*; under this view, information is data relevant for a particular decision. To illustrate this, take again the example of deciding where to advertise the new (and, no doubt, much improved!) bubblebath and assume that, in addition to the survey on reading habits mentioned above, you are also given the following: (a) a list of the circulations and advertising rates of different newspapers/magazines, (b) a list of print media used by your competitors, and (c) the shoe size of your boss's secretary. While most of us would agree that (a), (b) and (c) all constitute some kind of data, practically noone (sober, at least) would consider (c) as *information*. Of course, for a different decision, (c) may be a perfectly legitimate informational input (e.g. if the objective is to order sports shoes for the office rugby team).

Summary

Let us take a big breath and try to summarize what we have learned so far. We first looked at the nature of data, distinguishing between units of analysis, variables and values; we then put the three together to create a data matrix. Next, we warned against confusing questions with variables and discussed different kinds of data according to their meaning, source and time dimension. Finally, we considered the link between data and information. It is now time to make a cup of coffee and mentally prepare for the things to come.

Questions and Problems

1. What is the difference between a variable and a value?

2. Give two examples of different units of analysis and two examples of variables that could be used to characterize them.

3. What determines the size of a data matrix (i.e. the number of rows and columns)?

4. Give three examples of multi-response questions. How many variables would you need to capture the answers to them?

5. Distinguish between facts, awareness, intentions, opinions, and motives, and give an example of each. Why is it important to distinguish between these types of data?

6. What would you consider to be the main advantages (viz. disadvantages) of primary versus secondary data?

7. What type of data do you need in order to make inferences regarding change?

8. What is the key difference between trend data and panel data?

9. It has been argued that 'students often wonder what statistics is and why they should bother to study the subject'. What is your view on this?

10. What is your favourite dish (we like to get to know our audience)?

Notes

1. Reichman, WJ (1981) *Use and Abuse of Statistics*, p. 28. London: Pelican.

Further Reading

Diller, DC (1991) *Handbook of Research Design and Social Measurement*, 5th edn. London: Sage Publications. An invaluable general source of reference to practically all aspects of the research process; it covers every topic one can imagine and, if it doesn't, it tells you where to find out about it.

Jacob, H (1984) *Using Published Data: Errors and Remedies*. London: Sage Publications. Good discussion of the various problems potentially encountered when relying on secondary data, together with suggestions for overcoming them.

Spector, PE (1981) *Research Designs*. London: Sage Publications. A concise introduction to the different types of research design, with particular emphasis on the design of experimental studies.

Does *sampling* have a purpose other than providing employment for statisticians?

The Nature of Sampling

Having taught you all we know about data sets, we shall ignore your cries for mercy and proceed to do exactly the same with regard to sampling. Crudely speaking, a **sample** is a part of something larger, called a **population** (or 'universe'); the latter is the totality of entities in which we have an interest, i.e. the collection of individuals, objects or events about which we want to make inferences. For example, in Table 1.4 earlier, the population implied was 'all insomniacs living in Auchtermuchty, Scotland'. Other examples of populations are 'all countries on the planet Earth'; 'all record stores in the UK'; 'all strikes at Greek knitting factories during 1975–79'; and 'all vegetarian chiropodists based in Greater Manchester'. We can now define a sample as *a subset* of a given population. Going back again to Table 1.4, the sample consisted of 600 insomniacs living in Auchtermuchty, who were fortunate enough to have their purchasing behaviour studied. Possible samples for the other population examples mentioned above, are 'member-countries of NATO'; '28 record stores located anywhere in Somerset and Devon'; 'strikes at Greek knitting factories during the first six months of 1977'; and '13 vegetarian chiropodists with city centre practices in Manchester' (note that these are merely possible samples that *could* be drawn from the above populations, not necessarily *good* samples – an issue to be addressed later).

The rationale for sampling is really very simple: by checking out part of a whole we can say something about the whole. However, you may ask: But why bother with a sample in the first place? Why not study the population instead? Well, in some instances the entire population *is* studied as, for example, with the population **census** which is conducted every few years by the government. However, in most cases, undertaking a census is not possible or desirable for reasons summarized in Table 2.1; in fact, 'if a census is necessary to justify a conclusion, then the conclusion is probably not justified'.[1]

In spite of the many advantages of sampling over a census, it is important to keep in mind that 'by sampling, one buys reduced cost of data-collection, -processing and -analysis . . . but at the expense of adding one problem to the other problems offered by one's research: that of deciding whether the propositions established for the sample can be generalized to the universe'.[2] To illustrate this problem, we have applied our magnificent

sample
population

census

Reason	Rationale	Example
Cost	A census is almost always more expensive than taking a sample; sometimes the cost of a census is simply prohibitive.	Interviewing all car owners in all EU member-countries as to the colour of windscreen washer liquid they prefer would cost millions of pounds; the value of the information obtained to a car accessories manufacturer would not outweigh the cost of commissioning such a census.
Time	A census may take more time to conduct and analyse than is available for making the decision involved; in other words, doing a census may simply take *too* long.	The time needed to complete a census of all ski-enthusiasts in North America regarding their ski resort preferences would be unacceptably long if the research related to the updating of a *Snob Skiers Guide* which *must* hit the bookshops by September (and it's already May!).
Destruction/ contamination of population members	A complete census may result in destruction or contamination of the entire population; 'destructive testing' procedures must, by necessity, be limited to a sample only.	Torture-testing each and every condom coming off the production line for durability, tensile strength and friction resistance would leave no products reaching the shops (and a lot of pregnant people!).
Decision importance	A minor decision would not normally justify the hassle of a census; a 'quick and dirty' sample may be all that is needed.	Deciding where to place a cold drinks dispenser in a factory with 600 employees would not justify a census of 'perceived optimal locations'; on the other hand, whether to move the factory to a new location 25 miles away is an issue for which a census of the workforce's intentions to stay with the firm may be called for.
Confidentiality	A census is more likely to be noticed by interested parties than a sample study; with a census the chances that competition will get wind of what's going on are much higher.	Test-marketing a new family-size package of beetroot pickle by placing it in *all* outlets of the major supermarket chains (e.g. *Strangeway*, *Fresco*, and *Painsbury*), is much more likely to invite competitive retaliation than if only a few carefully selected stores are used for test-marketing purposes.
Accuracy	Unbelievable as it may sound, a census may be *less* accurate than a sample; the sampling error of the latter may be outweighed by the non-sampling error of the former resulting in greater *total* error in the case of a census.	Sending out 75 hastily recruited part-time interviewers to record the views of the entire student body of Edinburgh University on the introduction of a mandatory 'tartan dress code', and using another 20 semi-qualified, bored-to-death typists to input the data to the computer is likely to result in many non-sampling inaccuracies (e.g. interviewer variability and bias, and/or transcription and typing mistakes).

Table 2.1 Reasons for preferring a sample over a census

Figure 2.1   Population and sample.

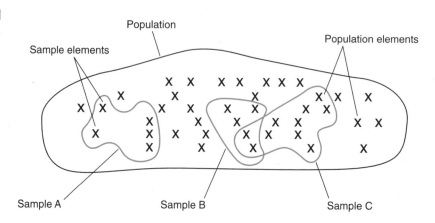

artistic talents and creatively depicted the relationship between population and sample in graphical form; the relevant masterpiece is shown in Figure 2.1.

 The first thing to realize is that, given a population containing a fixed number of elements (what is known among statistical geniuses as a **finite population**), there is a number of possible samples that could be drawn. Just to put it into perspective (and have you faint in the process), a total of 1 099 511 600 000 different samples could potentially be drawn from the population shown in Figure 2.1 (which only has 40 elements in the first place!). In general, if there are N elements in the population and the size of the sample is not fixed, one has a choice between 2^N possible samples (including taking no sample at all and conducting a census of the population). Of course, the choice of possible samples is drastically reduced if a certain **sample size** is pre-specified, that is, if the number of elements to be included in the sample is fixed beforehand (not an easy matter, either, as will be further discussed below). Going back to Figure 2.1, and assuming that we wish a sample of size 10 (such as sample C), our options are reduced to 'only' 847 660 530 possible samples! In general, given a desired sample size n (where $n \leq N$), then $\dfrac{N!}{n!(N-n)!}$ different samples could be drawn from a population with N elements. Thus, whichever way you look at it, one has a lot of choice when it comes to sampling.

Sample Selection

Given the above, the key question becomes how to select the n **sample elements** (i.e. members of the sample) so that they are representative of the N **population elements** (i.e. members of the population). Unfortunately, the answer to this is just as difficult as how to select a 'good' husband/wife (in other words, there is no easy answer!). One important consideration, however, is the effect that *excluded* population elements are likely to have on the 'quality' of the sample. Sampling, by definition, means that certain population elements will be excluded from the sample. This exclusion causes what is known as **sampling error**, that is, the difference between a result based on a sample and that which would have been obtained if the

Margin terms: finite population · sample size · sample elements · population elements · sampling error

entire population were studied (i.e. the 'true' value). Sampling error is generated whenever a sample is drawn by *whatever sampling procedure* and is a function of sample size (i.e. as the sample size increases, sampling error decreases).

To get a better feel of the concept of sampling error (while avoiding nasty technicalities), imagine that the population in Figure 2.1 consists of 30 female and 10 male students enrolled in an advanced entomology class at *Mosquito State University* and that you have just picked at *random* a sample of 5 students (Sample A); you could have done this by drawing names out of a hat or throwing darts on the class roster (blindfolded, of course). Now, it just *could* be the case that, as a result of pure chance, only male students were selected and, thus, Sample A would not be representa-, tive of the entire class (as female students – the majority in the population in this case – are not included in the sample). Remember, that you did not *intentionally* try to influence the composition of the sample – it just happened that no female students were chosen. However, the end result (i.e. no representation of female students) is exactly the same as if you had used a more 'subjective' procedure to select the sample, such as choosing only those people who are the same sex as you (assuming you are male) resulting in, say, Sample B.

So, is there a difference between Sample A (based upon random selection) and Sample B (based upon personal preference), if *both* contain sampling error and are both unrepresentative of the relevant population? Yes, there is. The difference is that with Sample A we can *assess* the extent of sampling error, whereas with Sample B we cannot. More specifically, Sample A was obtained by means of a **probability sampling** procedure, whereby each element in the population had a known, non-zero probability of being included in the sample (i.e. the sample elements were selected by chance and this chance was known for each element being selected), Consequently, we are able to apply the **laws of chance** (i.e. probability theory) and estimate how likely it is that it reflects the 'true' situation in the population. Sample B, on the other hand, was obtained by means of **non-probability sampling** procedure, since the selection of sampling elements was left to the discretion of the researcher and there was no explicit scientific model (such as probability theory) which could be used to assess the degree of sampling error.

Thus the key difference between probabilistic and non-probabilistic sampling methods is *not* that the former will always produce a more **representative sample** than the latter. Rather, it is that with the former, a *statistical evaluation* of sampling error can be undertaken, thus enabling the researcher to assess how *likely* the sample is to be unrepresentative and by how much; such an assessment is *not* possible with samples drawn by non-probabilistic methods (we will return to this topic in Chapter 9, where we will see how sample results can be used to estimate values in the population). It should also be borne in mind that **non-sampling errors** (e.g. measurement errors and non-response errors – see Chapters 3 and 4) also affect research results and, therefore, there is no guarantee that a probability sample will produce overall more accurate results than a non-probability sample.

probability sampling

laws of chance

non-probability sampling

representative sample

non-sampling errors

From the above, you may wonder why we should bother with non-probability sampling methods, given that no calculation of sampling error is possible. The reason for this is that, in addition to statistical criteria, practical considerations also influence the choice of sampling method. For example, applying probability sampling procedures tends to be more costly in time and/or money, requires a certain level of statistical training to use effectively, and assumes that a suitable **sampling frame** (i.e. a list of population elements) is available. Moreover, when one is not very concerned with the accuracy of **population estimates** (e.g. when only a 'first look', a 'rough and ready' or 'broad brush' view of the population is needed); when the population concerned is a **homogeneous population** (i.e. there is little variation among the population elements); and when the **expected cost of errors** in the obtained information is not very high (e.g. as in an unimportant decision), then using some kind of non-probability sampling procedure may be justified.

sampling frame

population estimates

homogeneous population
expected cost of errors

In terms of choice, there is a wide range of both probability and non-probability sampling methods available to suit all tastes, pockets, and desire for complicating things. Since there are many books on sampling (written by much better authors than us), we shall refrain from getting into the intricacies of the different methods but instead provide you with a quick summary of the most important ones (Table 2.2); those of you with a desire for punishment can look up the references at the end of the chapter for more details.

Table 2.2   Sampling methods

WARNING 2.1

Probability sampling procedures do *not* ensure that the sample will be representative or the results accurate. What they do, is allow the assessment of sampling error.

	Selection criteria	Example
Non-probability methods		
Convenience (chunk) sampling	Sample members are chosen on the basis of their being readily available/accessible; thus selection is done on the basis of convenience.	A sociology professor interested in people's attitudes towards mixed saunas, asks his introductory class to fill in a questionnaire on the subject.
Judgmental sampling	Sample members are chosen on the basis of the researcher's judgment as to what constitutes a representative sample for the population of interest; thus potential sample members are screened judgmentally as to whether or not they should be included in the sample.	A marketing researcher wishing to test-market the new *Starveline* range of diet foods uses his knowledge and expertise to select a sample of stores in key UK locations (e.g. John O'Groats, Llanerchymedd and Inverkirkaig) which are apparently reflecting national tastes.
Purposive sampling	Sample members are chosen with a specific purpose/objective in mind; the sample is thus intentionally selected to be non-representative.	A detergent manufacturer purposefully over-samples a number of current non-users of its washing powder to gauge reactions to a new formulation of the latter (the aim being to attract new customers while retaining existing ones).
Quota sampling	Sample members are chosen on the basis of satisfying some pre-specified criteria thought to apply to the population; the researcher is free to choose which elements to include in the sample as long as they qualify on the pre-defined characteristics.	A government department trying to assess the effectiveness of its 'Get A Job or Get Lost' advertising campaign (designed to promote employment opportunities) surveys 1000 adults of which 120 are unemployed and the rest in employment — the two groups (quotas) reflecting the current employment situation at national level.

Multiplicity (snowball) sampling	Sample members are initially chosen either judgmentally or through a probability sampling method and are subsequently asked to identify others with the desired characteristics; thus, the final sample is constructed from referrals provided by the initial respondents.	A professional association (e.g. *The National Society of Creative Accountants*) wishing to up its membership, contacts its existing members and solicits names of other individuals that may qualify for joining.
Probability methods		
Simple random sampling	Sample members are chosen randomly for inclusion in the sample, with each population element having an equal probability of being selected; further, each possible sample of *n* elements has a known and equal chance of being the one actually chosen.	*Excessive Premium Insurers*, an insurance company with 850 000 motor insurance policies randomly picks 2000 policy-holders and approaches them via direct mail to assess their level of satisfaction with the value-for-money provided.
Systematic sampling	Sample members are chosen at regular intervals after a random start; the sampling interval is the ratio N/n, where N and n represent the population and desired sample size, respectively.	A dubious car dealer with 3000 customers wants to survey 200 of them prior to offering a new 'all inclusive' warranty scheme; he thus sets a sampling interval of $3000/200 = 15$, picks at random a starting value between 1 and 15 inclusive – say, 8 – and subsequently selects from his customer records the 8th, 23rd, 38th and so on victim (sorry, customer!).
Stratified sampling	Sample members are chosen randomly from different *segments* (strata) of an overall population; each stratum may be sampled in proportion to its size in the overall population (proportionate stratified sampling) or sample members of different strata may have disproportionate chances of being selected (disproportionate stratified sampling).	*National Wetspinster Bank* splits its customer population into 'consumer' accounts (accounting for 80% of all accounts) and 'business accounts' (accounting for 20% of all accounts) and wants to conduct a survey of customer satisfaction among 500 account holders. If it wishes to retain the 80:20 consumer-to-business account ratio in the sample, then it would select at random 400 consumer accounts and 100 business accounts. If, however, for one reason or another, it is felt that greater variability in perceptions is likely to exist among business account holders, then more of the latter would be included in the sample; for example, the sample composition decided upon may consist of 300 consumer accounts and 200 business accounts (picked at random within each stratum).

Table 2.2 Sampling methods (continued)

Table 2.2 Sampling methods (continued)

Cluster sampling	Sample members are chosen in groups (clusters) rather than individually; the clusters themselves are chosen randomly from a population split into groups.	A sales director wishing to conduct a study of salesforce motivation, randomly picks 3 sales districts (out of a total of 8) and holds seven-hour, in-depth interviews with all 36 sales-people covering the chosen sales districts.
Multistage sampling	Final sample members are chosen by means of one of the other probability methods described above but a number of stages precedes the final selection.	*Exaggeration Research Incorporated,* an opinion poll organization, carrying out a nationwide poll on education standards, (a) selects at random 80 cities, then (b) selects randomly 30 blocks within each city, (c) takes a systematic sample of households within each block previously chosen, and (d) finally interviews the head of household (phew!).

One important point to note from Table 2.2 is that, in some instances, the sample elements (in the sense discussed previously) are not the same **sampling units** as the **sampling units**, the latter being the units actually chosen by the sampling procedure. More specifically, a sampling unit may contain one or more sample elements, as is the case, for example, with cluster sampling and multistage sampling. A major reason why sampling units and sampling elements may not coincide is that there may not be a suitable sampling frame available to enable direct selection of the latter. For example, if one wished to conduct a survey of marketing managers in the funeral industry, it might be difficult to obtain an appropriate register listing these individuals; as a result, one might have to sample funeral homes instead as the basic sampling unit (e.g. using the relevant trade association directory).

Sample Size Determination

Up to this point, very little attention has been paid to the question of sample *size* (other than pointing out that it is inversely related to sampling error). As was the case with the choice of sampling method, the determination of sample size is a rather complex issue, involving both statistical and practical considerations; we will only highlight key points in what follows.

A key statistical consideration in sample size determination is the **variability** degree of **variability** in the population; the more heterogeneous the population, the larger the sample size needed to capture the diversity in the population. For example, if the population consisted of 300 000 identical pink 40-watt lightbulbs, a sample of one would be sufficient to perfectly describe the population; on the other hand, if the population concerned consisted of the 40 entomology students of Figure 2.1, a larger sample would be needed (if only to capture both sexes, let alone differences in age, family background, etc.).

A second statistical consideration is the desired degree of **precision** associated with population estimates based on a sample; the greater the precision required, the larger the sample size needed. For example, if one wanted to estimate the average height of the 40 students in the entomology class within ± 1 cm, a larger sample would be needed – all other things being equal – than if one wanted an estimate within ± 10 cm of the true average height.

A third statistical consideration relates to the desired degree of **confidence** associated with any estimates made. Sticking to our brilliant entomology example, if one wished to estimate the average height of the students in the class with a precision of, say, ± 5 cm *and* wanted to be 95% confident that this estimate will contain the true population value, a larger sample size would be required than if one wanted to have an estimate of the same precision but a lower confidence level (e.g. 90%). It does not take a PhD in Incredibly Advanced Mathematical Statistics to realize that there is a trade-off between precision and confidence with a sample of fixed size; thus, in determining sample size one has to balance these two considerations against each other (we will have more to say about this trade-off in Chapter 8).

A final statistical consideration concerns the extent to which the intended analysis will involve the use of sub-samples for **cross-classification** purposes and/or the use of statistical techniques which assume a minimum sample size to produce meaningful results. For example, if you were only interested in an overall picture of the characteristics of the students in the entomology class a smaller sample size would be required than if you had plans to cross-tabulate, say, three income categories (e.g. 'stinking rich', 'moderately comfortable', and 'totally broke') with two sex categories. In the latter case, you would be setting up a table with $3 \times 2 = 6$ cells and you would obviously need to have sufficient observations in all of them if you want to draw inferences about the sub-populations involved; to drive this point home, if you take a sample of, say, 5 students it would be *impossible* to fill all cells (since there are 5 people to fit into 6 feasible slots). Although it is difficult to generalize from project to project (particularly in the light of the practical limitations discussed below), as a rough rule of thumb, 'the sample should be large enough so that there are 100 or more units in each category of the major breakdowns and a minimum of 20 to 50 in the minor breakdowns'.[3]

Moving away from statistical criteria to more practical concerns, issues of resource availability in terms of time, money and personnel available also have a (sometimes overriding) impact on the size of the sample. The poorer you are, the greater the time pressure to conduct the study, and the fewer the subordinates, friends and/or relatives you can convince to give you a hand, the less likely it is that you will be able to go for a large sample. Expectations regarding **non-response** (e.g. when using a mail questionnaire) also need to be incorporated in the determination of sample size, and so do expectations concerning the **value of information** provided by different size samples in relation to their costs. With regards to the latter, the **Bayesian approach** to sample size determination provides a formal procedure for selecting the sample size that maximizes the difference between the expected payoff of sample information and the estimated cost of

precision

confidence

cross-classification

HINT 2.1

Think about the most important cross-classifications you want to produce with your data *before* you decide on the sample size.

non-response

value of information

Bayesian approach

fixed sample
sequential sample

sampling (see Further Reading). In this context, rather than drawing a **fixed sample** (i.e. determining sample size *prior* to data collection), sometimes a **sequential sample** is preferred. Under this approach, if the results are not conclusive after a small sample is taken, more observations are made; if the increased sample size still does not furnish conclusive results, more population elements are added to the sample and so on (until the information obtained becomes sufficient to permit a conclusion).

The Sampling Process

As a final point, aimed primarily at sending you into the depths of depression, Table 2.3 outlines the stages of the sampling process and, more importantly, indicates the kind of errors that can occur at each stage. The moral of the story is that the quality of the sample on which subsequent analysis is based is a major determinant of the quality of your conclusions. No matter how clever you are in the use of statistical techniques, you simply *cannot* produce meaningful, credible, and generalizable results from a poor sample.

Table 2.3 The sampling process

Stage	Description	Potential errors
1. Define the population	Specification of the target population in terms of elements (e.g. teenagers), sampling units (e.g. households), extent (e.g. in London) and time (e.g. having lived in London for at least a year).	*Population specification error:* the target population is inappropriate for the problem at hand. For example, surveying only housewives regarding their criteria for choosing a car would ignore that husbands can play a (sometimes useful) role in influencing the decision.
2. Specify the sampling frame	Specification of the listing, directory or roster from which the sample will be chosen (e.g. the telephone directory); a sampling frame is essential if a probability sample is to be drawn.	*Frame error:* the chosen sampling frame may be inaccurate (i.e. capture other populations in addition to the one of interest) or incomplete (i.e. exclude some population members) or may include certain individual population elements more than once. For example, using a trade association list to draw a sample of motorcycle dealers will not cover those dealers that are not members of the association (because they are too 'dodgy' to be accepted or too mean to pay the relevant fee!).
3. Select the sampling method	Specification of whether a probability or non-probability approach will be applied to draw the sample and exactly how the sample members will be selected.	*Selection error:* a non-representative sample is obtained as a result of a non-probability sampling procedure. For example, in a survey of attitudes towards animals in which a quota sample according to, say, age and location is used, bias will be introduced by an interviewer who systematically avoids

Table 2.3 The sampling process (continued)

Stage	Description	Potential errors
		homes with cats, dogs, etc. (because (s)he hates/is afraid of all animals, including goldfish).
4. Determine sample size	Specification of the number of sample elements to be included in the final sample as well as the number of any intermediate sampling units (e.g. when a multi-stage sampling procedure is being followed).	*Small numbers error:* the sample is too small to permit the desired statistical analysis and enable meaningful generalizations for the population. For example, making a random selection of 8 individuals from the electoral roll and using their voting intentions to predict the outcome of the forthcoming General Election would not impress many people.
5. Draw the sample and collect the data	Specification of the operational procedures for the selection of sample members and carrying out the fieldwork (e.g. call every 3rd name in the telephone directory; if not in, call back in ½-hourly intervals; if refuses to answer, pick next name, etc.).	*Sampling error:* the results based upon the sample will practically always differ from those that would have been obtained had the entire population been studied instead; recall that only probability methods allow for an assessment of sampling error.

Non-response error: the researcher may fail to contact some members of the selected sample and/or some contacted members may fail to respond to all or some part of the research instrument. For example, an unscrupulous interviewer may make up the answers for sample members who happened to be out when (s)he called (thus avoiding the need for a call-back on another cold winter night) or a respondent might simply refuse to cooperate (because of tiredness, sensitivity of subject matter, or because he hates researchers). |

Summary

Let us summarize our journey into the wonderful world of sampling. We started by looking at the reasons for sampling and then spent some time exploring the link between a sample and a population. Next, we introduced the concept of sampling error and distinguished between probability and non-probability sampling methods. The influences bearing upon sample size determination were considered next and we concluded our journey with a look at the stages of the sampling process and the things that can go wrong during it. Are you ready for more?

Questions and Problems

1. Give five reasons for taking a sample instead of conducting a census; then, give five reasons for doing a census rather than taking a sample!

2. How would you define a 'population'? If you have a population with 10 elements, how many different samples of size 5 can you draw?

3. What is sampling error? What are non-sampling errors?

4. Do probability sampling methods *always* produce a more representative sample than non-probability methods?

5. What are the advantages (*viz.* disadvantages) of probability versus non-probability samples?

6. What are the key determinants of sample size?

7. Select one probability and one non-probability sampling method and give examples of research situations in which you would use them.

8. Is there a difference between a sampling element and a sampling unit?

9. Describe the stages of the sampling process and give one example of an error that can occur at each stage.

10. If you didn't have to learn about sampling, what would you be doing right now?

Notes

1. Peterson, RA (1982) *Marketing Research*, p. 333. Plano, TX: Business Publications Inc.
2. Galtung, J (1969) *Theory and Methods of Social Research*, p. 51. London: George Allen & Unwin.
3. Sudman, S (1976) *Applied Sampling*, p. 30. San Francisco: Academic Press.

Further Reading

Cochran, WG (1977) *Sampling Techniques*, 3rd edn. New York: Wiley. A classic on the subject, although very technical and heavy-going; an excellent source of reference, nevertheless.

Henry, GT (1990) *Practical Sampling*. London: Sage Publications. A practical, down-to-earth introduction to the everyday problems of sampling.

Sudman, S (1976) *Applied Sampling*. San Francisco: Academic Press. Another classic; eminently readable, full of examples and with a good discussion of the Bayesian approach to sampling.

3

Why should you be concerned about different types of *measurement* and are some better than others?

The Nature of Measurement

Having provided you with an unsurpassable discussion of sampling issues in the previous chapter, we need to complement this with an equally outstanding examination of measurement questions (yes, we know, modesty is one of our 328 virtues – shyness is another one!). In terms of the basic structure of the data matrix, sampling considerations relate to the *rows* of the matrix (i.e. the *units of analysis*), whereas measurement considerations relate to the *columns* of the matrix (i.e. the *variables* and the assignment of *values* to them). Incidentally, if you don't remember what a data matrix looks like, then go back to Chapter 1 and have another look at Table 1.2.

Before we can look at how to measure something, we must think a little about what we want to measure, i.e. we need to define the **concept** involved. Concepts express abstractions formed from observations from numerous particular happenings. For example, the concept 'bicycle' refers to the generalization of the characteristics that all bicycles have in common. In scientific research we also often speak of **constructs**; these are concepts that have been consciously and deliberately invented for particular scientific purposes (e.g. 'introversion', 'product life cycle', 'socialization'). For practical purposes, the terms 'concept' and 'construct' are often used interchangeably.

Defining a concept (or construct) is not always easy, particularly when the concept we are interested in does not have a physical referent. Compare, for example, the concepts 'bicycle' and 'brand loyalty'. The former is much easier to explicate as it is closely related to a physical reality and little disagreement would result if different people were asked to explain its meaning. In contrast, 'brand loyalty' is an abstraction that is much more difficult to define and, thus, measure.

In defining concepts, it is important to distinguish between two approaches. A **conceptual definition** defines a concept in terms of other concepts, the meaning of which is assumed to be more familiar to the reader. For example, brand loyalty has been conceptually defined as 'the preferential attitudinal and behavioral response towards one or more brands in a product category expressed over a period of time by a consumer (or buyer)'.[1] A conceptual (or 'constitutive') definition is roughly equivalent to a dictionary definition and aims to (a) capture the essence or key idea of the concept, and (b) distinguish it from other similar but, neverthe-

concept

construct

conceptual definition

operational definition

less, distinct concepts. An **operational definition**, on the other hand, describes the meaning of a concept through specifying the *procedures* or operations necessary to measure it. For example, brand loyalty may be operationalized as 'a sequence of consecutive purchases (typically, three or four) of the same brand'. Thus, an operational definition aims to translate the concept into observable events by specifying what the investigator must *do* in order to measure the concept concerned.

Clearly, a conceptual definition logically precedes an operational definition and, thus, it should be used to guide the development of the latter. In this context, it is possible – and often desirable – to have *multiple* operational definitions of the *same* concept. For example, brand loyalty could also be operationalized as 'the number of different (competing) brands purchased within a given time period', or 'the attitude of the buyer towards the brand'. Using multiple operationalizations is a good way to try to capture all the dimensions of a complex concept, thus allowing for comprehensive measurement. In our brand loyalty example, none of the operational definitions offered above captured both the attitudinal and behavioural aspects of the concept (although both aspects are equally relevant according to the conceptual definition). Thus, relying solely on any one of them would result in an incomplete picture at the data collection stage and, thus, an inadequate representation of the concept concerned.

It is important to appreciate that operationalization problems can be posed even by concepts that appear to be relatively clear and straightforward. Take, for example, the concept of 'firm size'. Although most of us would agree that IBM is a 'large' firm and the Chinese takeaway around the corner a 'small' business, if we wanted to take a sample of different-sized firms how should we actually define size? Should we look at sales volume (and, if so, at physical units or revenues)? Should we look at employment (and how do we deal with part-time employees)? Or should we look at value of assets (gross, net, including intangibles such as 'brand equity', or what)? What the complexities of this seemingly 'easy' operationalization task show is that 'in any measurement situation the measurement most obviously correct will ultimately turn out to be incorrect'.[2]

HINT 3.1

Use conceptual definitions as guides for developing operational definitions. Generate *multiple* operational definitions and employ a number of them to comprehensively represent your concept.

WARNING 3.1

There are *no* 'obvious' ways of operationalizing even the simplest concept. Think twice (and, then, think *again*!) before you decide on an operational definition.

Operational definitions and measurement go hand-in-hand, the former specifying how the latter should be undertaken for the concept(s) involved.

process of measurement

In general terms, the **process of measurement** can be thought of as the assignment of symbols to characteristics of persons, objects, states or events according to certain rules. There are three key points associated with this definition. Firstly, it is not the persons, objects, etc., themselves that are measured but rather their characteristics. For example, we do not measure our friend Rudolph, we measure his age, height, opinions, beer-drinking capacity, or some other characteristic; even when we count things, what we are in fact measuring is the characteristic of 'being present'.

measurement rules

Secondly, the assignment of symbols to characteristics is not done in an arbitrary manner, but according to pre-specified rules. **Measurement rules** ensure that the relations between the symbols assigned reflect the actual relations between the objects with respect to the characteristic concerned. To put it into statistical parlance (and give you an opportunity to practise your Greek), the assignment process is *isomorphic*, i.e. there is a one-to-one correspondence between the symbol and the characteristic being measured (otherwise the results of measurement would be useless since knowledge of a particular symbol would tell us nothing about the person or object concerned).

Lastly, in most measurement situations, the symbols assigned are *numbers*; the use of numbers rather than other symbols (e.g. letters, colours, etc.) provides a standardized means for communicating measurement procedures and results from researcher to researcher and user and also facilitates mathematical and statistical manipulation of the data. Thus, in practice, measurement can be seen as the use of numbers to represent the characteristics of persons, objects, events, etc. Note that the meaning of the term 'number' in a measurement context is *not* the same as that understood in everyday life, i.e. that numbers can be readily added, subtracted, multiplied and divided. While this *may* be the case, in measurement we can have different kinds of numbers, the precise meaning of which depends on (a) the nature of the characteristic being measured, and (b) the particular measurement rule used (indicating the basis upon which numbers are assigned to the characteristic).

Let us try and make this clear with an example. Suppose that we want to measure the weight of our beloved cousin Topsy. Although the obvious thing would be to ask/bribe/force Topsy to step on the scales and then take a reading (e.g. 174 kg), this is not the only measurement option available; Table 3.1 illustrates some alternatives.

Measurement scheme	Assignment of numbers
1. Create five categories	1=very heavy for her age 2=rather heavy for her age 3=roughly average for her age 4=rather light for her age 5=very light for her age
2. Compare Topsy to our other nine cousins	1=heaviest 2=2nd heaviest 3=3rd heaviest • • • 10=10th heaviest (i.e. lightest)
3. Define two classes	1=fantastic weight 2=horrible weight

Table 3.1   Measuring the weight of Cousin Topsy

It is evident from Table 3.1 that, depending upon the measurement scheme used, the meanings of the numbers differ. What this implies is that the same number can tell us different things and, therefore, if we do not know the measurement rule behind the number we cannot interpret it in a meaningful way.

WARNING 3.2

Numbers in measurement function as *symbols* — as such, they can be used in different ways and, thus, their meaning is *not* invariant.

Measurement Scales

Different measurement rules result in different types of **measurement**   **measurement scales**

scales. The latter can be visualized as a continuum upon which the measured objects, persons, etc., can be located. The measurement schemes in Table 3.1, for example, use different kinds of scales on which to place Topsy. A key distinction between different types of measurement scales is according to the **level of measurement** they provide, i.e. the amount of information they convey about the measured objects and the permissible mathematical/statistical operations that can be applied to the resulting data. In this context, four major types of measurement scales can be distinguished (see Table 3.2).

level of measurement

Table 3.2 Types of measurement scales

| Properties | Scale type | | | |
	Nominal	Ordinal	Interval	Ratio
Equivalence	Yes	Yes	Yes	Yes
Order	No	Yes	Yes	Yes
Equal intervals	No	No	Yes	Yes
Absolute zero	No	No	No	Yes
Typical usage	Store types	Occupation	Index numbers	Sales
	Product categories	Social class	Temperature	Costs
	Geographical	Brand preference	Calendar time	Age
	locations	Attitudes	Attitudes	Number of
				customers

nominal scale

A **nominal scale**, as the name implies, is a scale 'in name only' and represents the simplest type of scaling. In nominal scaling, the numbers used have no mathematical properties in themselves and serve only as labels for identification and/or classification. For example, assigning a unique number to each athlete in a sporting event serves only to identify the individual athletes taking part (otherwise how would you know who won?). This is the most elementary nominal scale (sometimes referred to as a **label nominal scale**); there is a strict *one-to-one* correspondence between each number and each athlete and, as long as this correspondence is preserved, any set of numbers could be assigned.

label nominal scale

category nominal scale

A more common nominal scale is the **category nominal scale**, whereby the numbers assigned represent *mutually exclusive* and *collectively exhaustive* categories of persons, objects, etc. For example, classifying individuals according to their nationality, eye colour or sex is done by means of category nominal scales. As long as the same number is not assigned to different objects or different numbers to the same object, again, any set of numbers can be used. For example, in classifying individuals according to nationality, the following schemes would furnish identical results: [1 = British, 2 = Korean, 3 = German]; [1 = Korean, 2 = German, 3 = British]; [–7 = German, 23 = Korean, 247 = British]; [456 = German, 0 = British, 9 = Korean]. This illustrates that the only property conveyed by the numbers in a nominal scale is that of **equivalence** (i.e. identity) and, consequently, the only legitimate mathematical operation is *counting*, i.e. the enumeration of persons, objects, etc. falling, in each category. For example, if we classified 300 individuals according to their nationality using any of the scales above, we could conclude that we had so many Brits (e.g. 120), so

equivalence

many Koreans (e.g. 80) and so many Germans (e.g. 100). We could also say that 40% of our sample consisted of Brits and 60% of Koreans and Germans. Finally, we could state that the **mode** (i.e. the most frequently occurring value) was 1 (first measurement scheme), 3 (second measurement scheme), 247 (third measurement scheme) or 0 (fourth measurement scheme), representing in all cases the British. There is hardly any more analysis we can do with the results, such as calculating an arithmetic average (what does 'average nationality' mean?).

    An **ordinal scale** establishes an *ordered relationship* between persons or objects being measured. In ordinal scaling, numbers are used to indicate whether a person, object etc., has more or less of a given characteristic than some other person or object. However, the numbers do not provide information as to *how much* more or less of the characteristic is possessed by the person or object concerned. For example, a consumer may be asked to try out and rank four brands of cornflakes (A, B, C and D) according to 'digestibility' as follows: 1 = 'most digestible', 2 = 'second most digestible', 3 = 'third most digestible', 4 = 'least digestible'. Suppose that the following results are obtained: 1 = B, 2 = D, 3 = A, 4 = C. Now, while we know that the order of preference is B, D, A and C, we do not know anything about the *differences* in digestibility among the four brands (e.g. we cannot tell whether, say, the difference between B and D is less than, equal to or greater than the difference between D and A). Thus as long as we preserve the ordering relationship, any set of numbers could have been used in our scale (e.g. 3 = 'most digestible', 48 = 'second most digestible, 49 = 'third most digestible', 1245 = 'least digestible'). Notice that the assignment of low numbers to indicate better performance is a matter of convention/convenience only, since when we rank things we tend to think in terms of 'first class', 'second class', etc. We can rank equally well by assigning low numbers to indicate poorer performance (e.g. as in 'first-class idiot'!). In short, given that the key property conveyed by the numbers in an ordinal scale is that of *order*, we can transform the scale any way we wish as long as the basic ordering is maintained. Put into disgustingly complicated jargon, this means that any strictly increasing (i.e. positive monotonic) transformation is permissible on ordinal scales.

    As you may have gathered, an ordinal scale subsumes the features of a nominal scale (since equivalent entities – i.e. persons, objects, etc. – receive the same rank) and, therefore, all the mathematical/statistical operations applicable to a nominal scale are also permissible on an ordinal scale. However, since ordinal scales incorporate order as an additional property, we can do more in terms of analysis with ordinal data than with nominal data (e.g. in addition to computing percentages and finding the mode, we could also calculate the **median**, that is, the value above which and below which half the scores lie).

    An **interval scale** possesses all the characteristics of an ordinal scale (i.e. equivalence and order) and, in addition, is characterized by **equality of intervals** between adjacent scale values. In interval scaling, the numbers used permit inferences to be made concerning the *extent of differences* that exist between the measured persons, objects, etc., with regard to a particular characteristic. The distances between the numbers correspond to the distances between the persons, objects etc., on the characteristic concerned. For example, when we measure temperature by the Fahrenheit

mode

ordinal scale

median
interval scale
equality of intervals

scale, we can say that the difference between 70°F and 50°F is the same as the difference between 40°F and 20°F and that both are twice the difference between 10°F and 0°F; however, we *cannot* say that 50°F is 'five times as hot' as 10°F (i.e. suggest a 5:1 ratio). The reason for this is that the zero point on the Fahrenheit scale is arbitrary and does not reflect the 'true' zero of the underlying characteristic (i.e. absence of heat). To make this point clear, let us convert the temperatures given above into the Celsius scale (also an interval scale), using the famous formula: $C=(5F-160)/9$ (very useful to know if you are a foreigner living in the United Kingdom or the United States). The corresponding temperatures for 50°F and 10°F are respectively 10°C and −12.2°C, which clearly do not stand in a 5:1 ratio. However, the ratios between the differences are still maintained (i.e. the difference between 21.1°C and 10°C is equal to the difference between 4.4°C and −6.7°C and both are twice the difference between −12.2°C and −17.7°C). Since an interval scale has an **arbitrary zero point** but the difference between any two adjacent scale points is equal to the difference between any other two adjacent scale points, we can assign any set of numbers to the scale so long as the intervality is preserved (as was done, for example, with the temperature scales above). In disgracefully incomprehensible technical terms, this means that any positive linear transformation (i.e. of the form $y=a+bx$, $b>0$) will preserve the properties of the scale (note, in this context, that the conversion formula from Fahrenheit into Celsius is such a transformation). Our horizons for analysis are further expanded with interval scales, in that, in addition to calculating the mode and median, the **mean** (i.e. the arithmetic average) can also be computed from the data (as can various indicators of dispersion or variability – you will learn all about them in Chapter 8).

arbitrary zero point

mean

Finally, a **ratio scale** has all the features of an interval scale (i.e. equivalence, order, equality of intervals) plus an **absolute zero point** (also known as 'true' or 'natural' zero). In ratio scaling, the numbers assigned enable comparisons to be made between the measured persons, objects, etc., in terms of *absolute magnitude* on a given characteristic; equal ratios between the scale values correspond to equal ratios among the persons or objects concerned (which is not the case with an interval scale). For example, we can measure the speed of a motorcycle in kilometres or miles per hour. On either scale, a reading of zero actually corresponds to absence of speed (i.e. the motorcycle is ststionary). Moreover, we can say that driving at 240 kph is twice as fast as at 120 kph (a ratio of 2:1) or, converting into miles, that 150 mph is twice as fast as 75 mph (also a 2:1 ratio). The only legitimate transformation on a ratio scale is one that changes the unit of measurement. In emphatically unintelligible terminology, this means that the properties of the ratio scale are preserved up to a positively proportionate transformation (i.e. of the form $y=cx$, $c>0$); as you might have guessed, the conversion formula from kilometres to miles per hour is of this nature (since 1 mile=1.609 kilometres, which is another useful formula for Continental Europeans lost in the United Kingdom or United States!). In terms of analysis, ratio scales offer little additional advantage over interval scales as the vast majority of analytical methods are equally applicable to both kinds of scale; while there are some statistical indicators which are only appropriate for ratio data, they are infrequently used in practice.

ratio scale
absolute zero point

From the above discussion (see also Table 3.2), it should be clear that the four types of measurement scales are *nested* within one another: as one progresses from a lower level of measurement to a higher one, the properties of the former are retained and additional properties are gained (thus allowing for more refined measurement and better analysis). This means that while one can always go back to a lower measurement level from a higher one (e.g. from ordinal to nominal), the reverse is *not* generally true: one cannot normally generate a higher level of measurement from a lower one (e.g. convert nominal data into interval). This asymmetry implies that the higher the level of measurement, the greater the resulting flexibility in terms of subsequent analysis. More specifically, with measures yielding **metric data** (i.e. interval or ratio) one can apply a set of analytical techniques known as **parametric statistics**, whereas with **non-metric data** (i.e. nominal and ordinal), only the less powerful **non-parametric** statistical techniques can be used (both types of techniques will be discussed from Chapter 7 onwards, so do not despair – yet!). Thus, given the choice, one should always choose metric measures over non-metric ones.

Some of you may feel uncomfortable with Hint 3.2 because some variables, by their very nature, offer little choice in terms of measurement level. In contrast to, say, weight or height which are obviously **quantitative variables**, characteristics such as geographical location or race are obviously **qualitative variables** and can only be represented by nominal scales. Similarly, preferences for, say, six Uruguayan heavy-metal guitarists may only be amenable to a ranking procedure (e.g. 1st = Jose 'Animal' Martinez, 2nd = Juan 'Mad Reptile' Perez, . . . , etc.), precluding quantification of the distances/differences in creative talent between the 'artists'. However, while the nature of the characteristic under consideration is an important constraint on the discretion we have in measurement, in most circumstances we do have at least *some* choice. To illustrate this, have a look at Table 3.3 which shows a very common way of measuring age of respondents in questionnaire studies (most of you will, no doubt, have come across this response format in one of the numerous silly forms you had to complete at one time or another).

HINT 3.2

Always aim for the *highest* measurement level possible. If feasible, obtain *metric* data from your measures.

metric data
parametric statistics
non-metric data
non-parametric

quantitative variables

qualitative variables

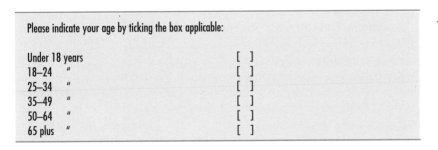

Table 3.3 Measuring age

Please indicate your age by ticking the box applicable:

Under 18 years	[ ]
18–24 "	[ ]
25–34 "	[ ]
35–49 "	[ ]
50–64 "	[ ]
65 plus "	[ ]

Being devastatingly perceptive, you will have noticed the wide variation in individual ages within each category (e.g. a 2-year-old still wetting his/her bed would be classified exactly the same as a 17-year-old about to enter university). You will also have noticed that there are unequal intervals across categories (e.g. compare the range of ages in the 18–24 category with that of the 35–49 category). Lastly, it would not have escaped your trained eye that there are open intervals (e.g. 65+), which means that we do not even know their true widths. All these problems are the direct result of the loss of information which has occurred because the measurement opportunities have not been fully capitalized. Age is, by nature, a ratio variable and, therefore, it could have been recorded by direct quantification (i.e. asking the respondent to state/write in his/her age in years). What we, in fact, have ended up with is an ordinal measure of a ratio variable and, consequently, we cannot even calculate the average age of the respondents, let alone make statements such as 'Respondent X is twice as old as respondent Y'. Put simply, we have thrown information away and there are no two ways about it.

Notice that, had we obtained the exact age of each respondent, we could (if we wished to) produce an age classification such as the one in Table 3.3 by appropriately grouping the individual ages (how this is best done will be discussed in Chapter 7). However, by virtue of our better measurement, we would also be in the position to comment on the average age of our sample, identify the oldest and youngest respondents, examine the proportion of respondents above/below any particular age, and so on.

Of course, if only to be awkward, you may argue that the real reason for using a measurement scheme such as the one in Table 3.3 is that respondents may be more willing to tick a box representing a range of ages rather than disclose their exact age (particularly in a face-to-face interview situation). This can certainly be true (as, for example, in the case of a great-grandmother enjoying her fourth youth) and can pose a response problem whenever 'sensitive' characteristics are to be measured (income, number of visits to disreputable night clubs in Amsterdam, tax bills, etc.). Under these circumstances one may consciously opt for a lower level of measurement than would have been ideal. Nevertheless, one must think very carefully whether the expected gain in the response rate actually outweighs the loss of information due to poorer measurement. Moreover, in studies promising anonymous/confidential treatment of responses and/or conducted by means other than personal interviews (e.g. mail surveys), there is much less reluctance to provide exact figures to open-ended questions, even for potentially sensitive variables. In short, Hint 3.2 stands strong and you should keep it in mind.

Sometimes, you can improve upon the level of measurement by creating another variable which is related to the one on which you have data. To do this you do not have to be the most creative person on earth (although this would help), but you definitely need an inquisitive mind and sufficient motivation (because there is additional work involved!). Imagine, for example, that you have just completed an international survey of mercenaries' attitudes towards disarmament in which, among other things, you have recorded the respondents' nationality on a nominal scale (e.g. 1 = Irish, 2 = Estonian, 3 = Peruvian, . . . ,47 = Greek, etc.). One thing you could do with this data is create another variable reflecting the level of economic development of each country concerned. This would involve the allocation of a new set of values to the various countries (e.g. 1 = post-industrial, 2 = industrial, 3 = developing, 4 = underdeveloped) and the resulting data could be treated as *ordinal* (reflecting successive levels of economic development). Of course, you would need to find out whether, say, Estonia, should be classified under 'developing' or 'underdeveloped countries', but a quick trip to your local library should take care of that (to consult appropriate secondary sources such as publications by the World Bank or the International Monetary Fund). Similar opportunities for generating new variables with a higher level of measurement exist for such characteristics as occupation (one can place individual occupations on an ordinal scale indicating occupational status or prestige, e.g. 1 = professional/managerial, 2 = skilled worker, 3 = unskilled worker) and education (one can fit individual qualifications according to educational level, e.g. 1 = postgraduate university education, 2 = undergraduate university education, . . . ,7 = part-time kindergarten only).

A common problem encountered with scales measuring characteristics such as attitudes and opinions is that the level of measurement implied is not always unambiguously clear (indeed sometimes it is unambiguously *un*clear). In most cases, this problem boils down to deciding whether the scale concerned yields only ordinal data or whether an assumption of intervality can be justified. Take, for instance, the scales majestically displayed in Table 3.4, all of which attempt to measure people's perceptions towards Dan Quayle's US vice-presidency.

HINT 3.3

Investigate opportunities for improving upon the measurement level through the creation of new variables. Be creative!

Table 3.4   Dan Quayle's vice-presidency

A. *Dan Quayle has been an excellent vice-president.*				
Strongly agree	Agree	Neither agree nor disagree	Disagree	Strongly disagree
[ ]	[ ]	[ ]	[ ]	[ ]

B. *Dan Quayle's vice-presidency has been*						
Excellent [ ]	[ ]	[ ]	[ ]	[ ]	[ ] Appalling

C. *As a vice-president, Dan Quayle has been*				
Very competent	Somewhat competent	Neither competent nor incompetent	Somewhat incompetent	Very incompetent
[ ]	[ ]	[ ]	[ ]	[ ]

D. *Vice-president Dan Quayle has been*
+3
+2
+1
competent
−1
−2
−3

The scales in Table 3.4 are all examples of the most widely used attitude-measurement techniques in social and marketing research (A is a **Likert scale**; B is an example of the **semantic differential**; C is an **itemized rating scale**; and D is a **Stapel scale**). Are these scales interval or ordinal? Unfortunately, to answer this question properly we would have to spend at least three years delving into the relevant (and horrendously extensive) literature. Since we feel that you would not want us to do this, we shall settle this matter by first observing that 'if we use ordinal measures as though they were interval, we *can* err seriously in interpreting data

Likert scale
semantic differential
itemized rating scale
Stapel scale

and the relations inferred from data . . . . On the other hand, if we abide strictly by the rules, we cut off powerful modes of measurement and analysis and are left with tools inadequate to solve the problems we want to solve'.[3]

Bearing these words of infinite wisdom in mind, if we adopt the 'pragmatic' view followed by most social researchers, then we would treat the scales in Table 3.4 (or similar ones) *as if* they were interval. It is recommended, in this context, to appropriately number the response alternatives on the scale so as to communicate to the respondent that the intervals between the scale points are intended to be of equal distance. On the other hand, if we adopt the 'purist' view most commonly followed by statisticians, then the scales of Table 3.4 should be treated as ordinal (unless we can *prove* otherwise).

Just so you don't get carried away, if you opt for the pragmatic view, you must be alert to the possibility of grossly unequal intervals and be cautious with the interpretation of data that are assumed to be interval. Moreover, you will *not* get away with it if you adopt the pragmatic view for scales which are patently ordinal. For example, if a respondent is asked to compare brand A of dog shampoo with brand B, on a scale reading 1 = worse, 2 = about the same, 3 = better, the resulting data cannot be reasonably treated as interval (so don't try tricks like that, or you'll be in trouble!).

At this point, we must deal briefly with a special kind of measurement situation, namely that involving **dichotomous variables**. A dichotomous (or 'binary') variable takes only two values (e.g. male/female, user/non-user, pass/fail) and is typically scored via **dummy-variable coding**, i.e. by assigning the value of 1 to one category (usually to mark the presence of some characteristic) and 0 to the other category (to mark its absence). At first glance, dichotomous variables appear to reflect a nominal scale; for example, sex has two unordered categories (male/female) implying nominal level of measurement. However, a dichotomous variable can also be viewed as reflecting both an order as well as equal intervals (see Table 3.2). Concerning the former, although a 'natural' rank order may not be inherent in the category definitions (e.g. as in the case of sex), either arrangement of the categories satisfies the mathematical properties of ordering. It makes no difference whatsoever which end of the ranking is considered 'high' or 'low' (thus 1=female, 0=male and 1=male, 0=female are equally acceptable). Moreover, the requirement of equally-sized intervals is also satisfied because there is only one interval (which is naturally equal to itself). In short, dichotomous variables can be treated as nominal, ordinal or even interval, depending upon the research situation – isn't this great?

WARNING 3.3

Think *very* carefullly whether the scale you are using can be assumed to be interval for analysis purposes. Think *even more* carefully about how you will interpret the results.

dichotomous variables

dummy-variable coding

Scaling Formats

scaling formats

Let us attempt to confuse you even further by having a quick look at the bewildering variety of scaling techniques at your disposal. Table 3.5 summarizes the main **scaling formats** you are most likely to encounter in data analysis projects.

As you can see, in many cases there is considerable scope for 'customizing' or 'tailoring' the final form of your scale by manipulating one or more

Table 3.5 Different kinds of scaling formats

Format	Basic form	Description	Example
Checklist	Chose *m* out of *n*	Repondent is given *n* options and asked to select up to *m* ($m \leq n$).	*Which of the following items do you own?* 1. A car 2. A cattle prod 3. A yellow tea-cosy 4. A pair of purple wellies
Determinant choice	Choose 1 out of *n*	Respondent is given *n* options and asked to select one of them.	*Which of the following persons should be the next Prime Minister?* 1. John Major 2. John Travolta 3. John 'spoiled brat' McEnroe
Dichotomy	Yes/no	Respondent is asked to respond with a 'yes/no' answer.	*Have you ever visited Siberia?* 1. Yes 2. No
Rank Order	Rank *n*	Respondent is given *n* options and asked to rank them in ascending or descending order.	*Please rank the following chocolate bars according to taste (1 = best, 4 = worst)* Supernut _____ Thunderslice _____ Almondchunk _____ Crumbleflake _____
Constant sum	Allocate *n* to *m*	Respondent is given *m* attributes and asked to allocate *n* points (usually 100) to them.	*Please allocate 100 points to the following attributes according to their importance in your choice of pub.* Beer ____ Barmaid ____ Clientele ____ Opportunity for fights ____ Total: 100
Paired comparison	Choose X or Y	Respondent is given two objects (X,Y) and asked to select one according to the characteristic in question.	*Which of the following brands of soap would you choose if both cost £3.95 per bar?* 1. Moroccan Sunset 2. Andalusian Mist
Comparative rating	Rate X against Y on *n*	Respondent is given a rating scale with *n* points and asked to compare object X against Y by selecting one point on the scale.	*In terms of speed, how does the new Fjord Siesta compare with the new Tinnie Retro?:* 5. Much faster 4. Somewhat faster 3. About the same 2. Somewhat slower 1. Much slower

Table 3.5 Different kinds of scaling formats (continued)

Format	Basic form	Description	Example
Direct rating	Rate X on *n*	Respondent is given a scale with *n* points and asked to rate object X by selecting one point on the scale.	*Having just driven the new Fjord Siesta how would you rate its windscreen wiper speed?:* 5. Extremely fast 4. Very fast 3. Average 2. Very slow 1. Extremely slow
Direct quantification	Give a figure	Respondent is asked to state (or write in) a fact that can be expressed as a number (integer or real).	*On average, how many pints of Czechlager do you drink per week?* ☐☐☐pints

of its basic features. To illustrate the opportunities (read: complexities) involved, consider the direct rating scale. Among the decisions you will have to make in using such a scale, are (1) whether to provide an *odd v. even* number of response alternatives (i.e. whether to include a middle/neutral point), (2) whether to use a *balanced v. unbalanced* distribution of response alternatives (i.e. whether to provide an equal number of favourable/positive and unfavourable/negative scale points), (3) whether to use a *forced v. unforced* response format (i.e. whether to provide a 'no opinion', 'not applicable' or 'no knowledge' category), (4) whether to label *all v. some* of the scale points (i.e. whether to provide a verbal description of each response alternative or label only the extremes and/or middle point on the scale), and (5) *how many* response alternatives (i.e. scale points) to include. Different research problems, respondent groups and characteristics to be measured all affect the precise form of the scaling format finally chosen. While space limitations preclude a discussion of the individual impact of these factors on (1)–(5) above, the incredibly stimulating list of references we have generously provided (see Further Reading) contains comprehensive treatments of these issues.

Measurement Error

measurement quality

A final but very important issue that we need to deal with before we say goodbye to measurement concerns **measurement quality**. In an ideal world, every time we measured something we could be sure that the score (i.e. value) obtained actually reflected the 'true' value of the underlying characteristic *and nothing else*. In other words, there would be a perfect one-to-one correspondence between the **observed score** (O) (i.e. our measurement) and the **true score** (T) (i.e. the reflection of the characteristic of interest) and, consequently, **measurement error** (E) would be zero.

observed score
true score
measurement error

Unfortunately, since we do not live in a perfect world (the atrocious weather in Britain is indisputable evidence of this), measurement error is not normally zero and, therefore, we cannot take for granted that our

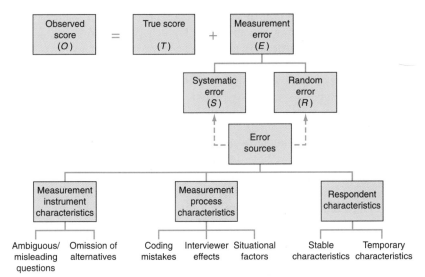

Figure 3.1   Measurement in the real world.

measures in fact do a good job. Figure 3.1 shows our predicament by highlighting the different components of measurements and the sources of measurement error.

What Figure 3.1 tells us is that, given half a chance, measurement error will creep in whenever we are administering a measurement instrument. Respondent characteristics, both stable (e.g. respondent likes to disagree 'just for the hell of it') and temporary (e.g. respondent is in a bad mood), together with imperfections of the measurement instrument (e.g. lack of a 'don't know' alternative on a scale presented to an undecided respondent) and problems of the measurement situation (e.g. interviewer is fancied by the respondent while extremely jealous husband/wife is present), all combine to make our life difficult by introducing error in our measurements.

Two types of error can be distinguished in this respect. **Systematic error** (S) (also known as 'bias') occurs in a consistent manner every time a measurement is taken (e.g. a general tendency to respond negatively independently of one's true feelings, as in the case of the 'professional disagreer' above). Thus systematic error results in either inflation (overestimation) or attenuation (underestimation) of the true score. **Random error** (R) (also known as 'variable error'), on the other hand, does not manifest itself consistently every time a measurement is taken (e.g. a temporary characteristic such as mood may be reflected in overly favourable responses if the respondent is in a good mood and in excessively negative responses if the respondent is in a foul mood). Thus, random error tends to be self-compensating since it can occur in either direction.

The extent to which a particular measure is free from both systematic and random error indicates the **validity** of the measure; a perfectly valid measure implies O=T (i.e. the measurement obtained reflects only the true score on the characteristic of interest). The extent to which a measure is free from random error indicates the **reliability** of the measure; a perfectly reliable measure implies R=0 (i.e. no random error component). It can be seen that reliability is a necessary but not sufficient condition for validity (since, even if R=0, the measurement obtained could still contain

systematic error

random error

validity

reliability

systematic error in addition to the true score, i.e. $O=T+S$). If a measure is not reliable then it cannot be valid, but if it is reliable it may or may not be valid; put differently, a measure that is valid is also reliable but the reverse is not necessarily true. Confused? Don't worry, so are we! A rather annoying detail in any practical attempt to assess the validity or reliability of a measure is that we do not *know* the true score of the respondent on the characteristic concerned (if we did, there would be no point in measuring it and we could all go home and sleep!). Therefore, we must somehow try to gather some sort of evidence that will enable us to *infer* the extent to which our instrument is valid and reliable; yes, this does involve even more work, but that's science for you!

In validity assessment, the basic question that we try (usually unsuccessfully!) to answer is: 'Are we in fact measuring what we think we are measuring?' Thus our intention is to show that our measurement device does in fact measure what it appears to measure. You may wonder why we are asking what appears to be a rather trivial question. For example, if we employ a ruler to measure our desk, it is pretty obvious that what we are measuring is size (i.e. length/width/height) and not sex appeal or religious beliefs. This may be so, but what if we are using a certain scale to measure 'citizen satisfaction' (a feeling) with 'police services' (an intangible and multifaceted object of evaluation)? Is our scale really a satisfaction scale or is it capturing other properties such as 'discontent', 'alienation' or 'law-abidingness' (got you there, eh?). Even with measures which are 'obviously' measuring 'obvious' things (such as with our ruler example above), problems of validity may still crop up; thus, a poorly calibrated ruler will provide very precise but wholly inaccurate (and, hence, invalid) measurements owing to the presence of systematic error. Thus, attention to establishing the validity of one's measures is good research practice and time well spent. The bad news is that validity assessment can be quite complex,

Table 3.6 Validity assessment approaches

Approach	Description	Procedure
1. Content validity	The extent to which a measure *appears* to measure the characteristic it is supposed to measure.	Subjective assessment of the appropriateness of the measure for the task at hand.
1.1 Face validity	The extent to which a measure seems to capture the characteristic of interest.	Agreement between expert and/or non-expert judges as to the suitability of the measure.
1.2 Sampling validity	The extent to which a 'content population' of situations/behaviours relating to the characteristic of interest (i.e. the characteristic's conceptual domain) is adequately represented by the measure concerned.	As above

Approach	Description	Procedure
2. Criterion validity[a]	The extent to which a measure can be used to predict an individual's score on some other characteristic (the criterion).	Examination of the relationship between the measure and a criterion.
2.1 Concurrent validity	The extent to which a measure is related to another measure (the criterion) when both are measured at the *same* point in time.	Comparison of the scores obtained on the measure concerned and those obtained on the criterion
2.2 Predictive validity	The extent to which *current* scores on a given measure can predict *future* scores of another measure (the criterion).	As above
3. Construct validity	The extent to which a measure behaves in a theoretically sound manner.	Investigation of the relationships between the measure concerned and measures of other concepts/characteristics within a theoretical framework.
3.1 Convergent validity	The extent to which a measure is positively related to other measures of the same concept obtained by independent methods.	Examination of the relationships between measures of the same concept generated by different methods.
3.2 Discriminant validity	The extent to which a measure is not related to measures of different concepts with which no theoretical relationships are expected.	Examination of the relationships between measures of different concepts that are theoretically unrelated.
3.3 Nomological validity	The extent to which a measure is related to measures of other concepts in a manner consistent with theoretical expectations.	Examination of the relationships between measures of different concepts that are theoretically related.

Table 3.6 Validity assessment approaches (continued)

[a]Also known as pragmatic or empirical validity.

as there are a number of different angles one can take (see Table 3.6). The *really* bad news is that it is the collective picture painted by evidence relating to the various kinds of validity that determines the overall validity of a measure. Using our standard excuse of 'space limitations', we will not get into the nitty-gritty of validity assessment procedures but direct you, once again, to the Further Reading section for some excellent guides to the steps involved.

Shifting attention to the assessment of reliability, things are not as complicated as with validity assessment (but, then again, they aren't simple either!). The key question we are concerned with here is: 'Are we getting consistent results from our measures?' There are two types of consistency we are particularly interested in. The first is consistency over time, that is the extent to which we get similar results from repeated applications of the same (or similar) measurement instrument to the same set of

Table 3.7 Reliability
assessment procedures

Approach	Description	Procedure
1. Test–retest reliability	Consistency of results of repeated applications to the same measure to the same respondents.	Assessment of the degree of correspondence between two (or more) applications of the same measure under similar conditions.
2. Alternative forms reliability	Consistency of results of applications of 'equivalent' forms of the measure to the same respondents.	Assessment of the degree of correspondence between two equivalent measures administered under similar conditions.
3. Split-sample reliability	Consistency of the results of applications of the same measure across randomly selected sub-samples of respondents.	Assessment of the degree of correspondence between random sub-samples of respondents (usually split 50:50) on the same measure.
4. Internal consistency reliability	Consistency of the results across individual items comprising a composite scale.	Assessment of the degree of consistency within a multi-item measure administered to a group of respondents.
5. Scorer reliability	Consistency of the results provided of different judges or scores when asked to categorize open-ended responses.	Assessment of the degree of agreement between independent categorizations of a set of items by multiple judges.

stability

equivalence

respondents; this is known as the **stability** aspect of reliability. The second type of consistency is known as **equivalence** and indicates the extent to which the same set of respondents replies in a consistent manner on similar items; alternatively, equivalence can be seen as the extent to which different (but comparable) sets of respondents produce similar results on the same measurement instrument. Table 3.7 summarizes the key methods for assessing reliability in terms of stability and equivalence; detailed guidelines on how to go about it can be found (yes, you've guessed it) in the Further Reading section.

As if the notions of validity and reliability were not in themselves complicated enough to drive any sane person insane, two additional dimensions are sometimes referred to in discussions of measurement quality.

sensitivity

Sensitivity refers to the extent to which a particular measure is able to capture variability in responses and is a particularly desirable property when *changes* in the characteristic of interest are measured. For example, a dichotomous scale such as 1=agree, 0=disagree is unlikely to capture subtle attitude changes to the same extent as a 5-point scale anchored at 5=strongly agree, 1=strongly disagree and also containing a middle point.

generalizability

Generalizability refers to the extent to which a scale is applicable and interpretable in different research settings. For example, is a scale intended to measure 'hedonism' easy to employ with different data collection methods (e.g. mail v. telephone interviews), equally applicable to different respondent groups (e.g. males v. females) and readily interpretable in different situations (e.g. in single-country v. cross-cultural investigations)? In one sense, sensitivity and generalizability can be seen as sub-dimensions of validity and reliability. Thus if a scale developed in setting X cannot be applied to setting Y (i.e. is not generalizable), this is because *for setting Y* the scale has low reliability and/or low validity. Equally, a scale exhibiting low sensitivity is unlikely to produce impressive results in reliability or validity assessments.

Summary

In this mind-blowing chapter we looked at a variety of issues relating to measurement and scaling. We began by considering the notion of measurement and the rationale behind the measurement process. We then distinguished between conceptual and operational definitions and followed this with a discussion of the different types of measurement scales. Next, we tried to make some sense out of the various scaling formats and called it a day with a brief look at measurement error, emphasizing validity and reliability considerations. If you need a drink, now is the time to get it.

Questions and Problems

1. Why do we need conceptual definitions?

2. What is the role of numbers in measurement?

3. Describe the different levels of measurement and give one example of a measurement scale for each.

4. What is the difference between an interval and a ratio scale?

5. Why should you generally aim for the highest level of measurement possible?

6. What is so special about dichotomous variables?

7. What are the key factors determining the final format of a rating scale?

8. Which is more important in measurement: validity or reliability?

9. Give three examples of sources of measurement error; then, give three more!

10. How would you go about developing a measure of the 'effectiveness' of this book in teaching you data analysis?

Notes

1. Engel, JF & Blackwell, RD (1982) *Consumer Behaviour*, 4th edn, p. 570. Hinsdale, IL: Dryden Press.
2. Peterson, RA (1982) *Marketing Research*, p. 261. Plano, TX: Business Publications.
3. Kerlinger, FN (1964) *Foundations of Behavioral Research*, p. 427. London & New York: Holt, Rinehart & Winston.

Further Reading

Carmines, EG & Zeller, RA (1979) *Reliability and Validity Assessment*. London: Sage Publications. A very readable introduction to different approaches for evaluating the validity and reliability of measures.

Oppenheim, AN (1966) *Questionnaire Design and Attitude Measurement*. London: Heinemann. A classic and very easy to follow text on measurement and scaling, emphasizing attitude measurement problems.

Spector, PE (1992) *Summated Rating Scale Construction: An Introduction*. London: Sage Publications. If you need to develop a scale from scratch, you cannot do better than start here.

DeVellis, RF (1991) *Scale Development: Theories and Applications*. London: Sage Publications. Read this one after you've read Spector (1992) to get the maximum benefit. Good stuff.

PART II:

PREPARING DATA FOR ANALYSIS

4

Have you cleaned your data and found the *mistakes* you made?

The Role of Data Cleaning

In Chapter 1 we pointed out that the raw material for analysis is always a data matrix, the rows of which reflect the units of analysis and the columns the variables on which the units of analysis are measured. In Chapters 2 and 3 we elaborated on the basic structure of the data matrix by discussing sampling issues (affecting its rows) and measurement issues (affecting its columns). Figure 4.1 summarizes what we have covered so far and also indicates the next two stages involved, notably (a) preparing the data for analysis by means of editing and coding and (b) transforming the data into *results*.

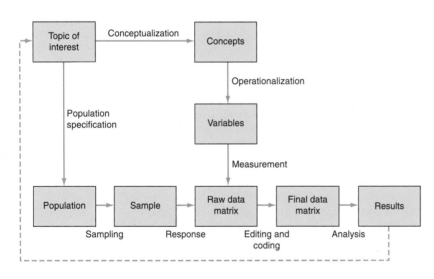

Figure 4.1 The long journey to producing results.

In this and the next chapter we will deal with the preparation of data for analysis, i.e. how one goes about 'cleaning' and coding data and what kind of computer packages one may choose for the actual analysis.

To avoid a possible misunderstanding right away, **data cleaning** does not mean you have to wash your completed questionnaires, take showers with your data matrix or make sure that no obscene words slip into the responses to your open-ended questions! Instead, the objective is to

data cleaning

'identify omissions, ambiguities, and errors in the responses'.[1] Thus data cleaning aims at avoiding errors in the data matrix questionnaires both during and immediately after the collection of your data. Usually, this process is referred to as **editing** and comprises several aspects, as shown in Figure 4.2.

editing

Figure 4.2   *Main editing tasks.*

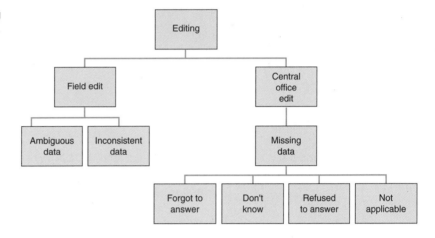

field edit

The **field edit** is primarily concerned with controlling interviewers through monitoring and validation procedures; the basic motto being 'trust is good, control is better'! Thus, if you are in the lucky situation to have people doing the data collection for you, you may want to re-contact a sample of respondents to ensure that the interviews actually took place and the answers were not made up by the interviewer while watching *Coronation Street*. To apportion appropriate blame, it is necessary to record the **interviewer number** on each questionnaire and also to assign a **respondent number**. This procedure also enables you to go back to the appropriate interviewer and/or respondent if you cannot read their handwriting or want to clarify confusing or contradictory responses (for example, if a respondent completed your open-ended question in Hindu and yours is a bit rusty these days).

interviewer number
respondent number

central office edit

The **central office edit** is only a second-best alternative to a field edit in that you now have to deal with ambiguous, inconsistent or missing data *without* (usually) being able to clarify the problem with the appropriate respondent. Consider the question 'When did you purchase your last car?'. During the editing process, you may find that the majority of respondents put the year (e.g. 1994) in the space provided. Some, however, may have answered the question in terms of months (e.g. 3 months ago) and one respondent from Huddersfield West answered 'When my third wife walked out on me and took my old car'. What, if anything, can we learn from this (apart from the fact that one has to be careful with giving a second car key to the wife)?

Firstly, that you can reduce the amount of editing work due to **ambiguous answers** by not asking ambiguous *questions*. For example, putting 'months' in brackets would have helped with our Huddersfield West respondent.

ambiguous answers

HINT 4.1

Always assign a respondent (questionnaire) number and use an interviewer code where a number of interviewers are involved in data collection.

Secondly, that you need to guard against creating confusion by using figures which may be ambiguous. For example, taking the average of figures partly relating to months and partly to years is unlikely to result in anything useful.

Thirdly, that you often need to make a pretty drastic decision about whether (a) the entire questionnaire ought to be thrown out (for example, if the respondent from Huddersfield West had answered the majority of questions in a similarly uninformative manner), or (b) a particular question should be ignored in further analysis (such as our example question above).

Related to the problem of ambiguous answers is that of **logically inconsistent data**. An extreme case would be a male respondent who proceeds to answer a series of questions relating to problems encountered during his last pregnancy! In such cases, you have to make certain that this data is not included in the analysis. However, you have to be careful with *seemingly* inconsistent data which are actually feasible. Consider a taste test between three new desserts: *Slimmo-Stop*, *Obese-Plus* and *Gut-Plug*. A respondent may prefer *Slimmo-Stop* compared to *Obese-Plus* and *Obese-Plus* compared to *Gut-Plug*. However, the same respondent may still prefer the taste of *Gut-Plug* compared to that of *Slimmo-Stop*. Thus, in general, responses of the nature A > B, B > C and C > A are sometimes feasible and are known as **intransitive responses**.

Perhaps the most common problem encountered when questionnaires are returned is that of **missing data**, which enables us to introduce yet another vital piece of jargon (useful as a conversation stopper when talking to your elderly aunt Gwen from Mid-Wales). The jargon is **item non-response** and refers to specific questions which have been left unanswered. When editing, you must make up your mind *why* they have not been answered and what to do about it. Usually, there are four possibilities why a question has not been answered:

(a) *The question did not apply to the respondent.* This may occur when the respondent is a left-handed tea drinker whereas the branching instructions in the questionnaire only required right-handed coffee drinkers to answer a particular set of questions.

(b) *The respondent refused to answer the question.* For example, refusals are to be expected when clergymen are asked whether they harbour any impure desires relating to professional rugby players.

(c) *The respondent did not know the answer to the question.* For example, most respondents would not know the cholesterol level of their dog.

(d) *The respondent simply forgot to answer the question.*

Remember that it is up to you to decide how best to deal with unanswered questions. In this context, you should consider whether it is possible or, indeed, necessary to distinguish between the above four reasons for item non-response. If such distinctions are important for your intended analysis, you should use *different* **missing values** (also known as 'missing data codes') for 'not applicable' (NA), 'refused' (RF), 'did not know' (DK) and 'forgot to answer' (FA).

logically inconsistent data

intransitive responses

missing data

item non-response

WARNING 4.1

Apparently logically inconsistent data (intransitive responses) Should not always be discarded as respondent error; especially in taste comparisons and importance rankings, they are possible.

missing values

Table 4.1 Measuring consumer reactions to a new chocolate spread

Having tasted our fantastic new chocolate spread, which one of the following words describes the taste of our excellent product most accurately?	
ecstatic	☐
outstanding	☐
exuberant	☐
thrilling	☐
mind blowing	☐

HINT 4.2

When it is not required to make a distinction between different reasons for item non-response, define only one missing value. However, make sure that this value lies outside the range of possible answers.

system missing value

WARNING 4.2

Do not replace an excessive amount of missing data with averages. It is both misleading and unethical.

Consider the example in Table 4.1 in which consumer perceptions to a new chocolate spread are measured. Say that some respondents ticked none of the response alternatives. If the question was supposed to be answered by all respondents, you may decide to interpret this as 'forgot to answer'. However, if the data you are editing is based on an interview with a respondent who, despite having received 27 complimentary packages, still could not bring himself to try your new chocolate spread, you may treat this as 'don't know'. Alternatively, a respondent may have scribbled on the questionnaire that she finds the chocolate spread truly repulsive and does not think the provided alternatives are fair; clearly a case of 'refused'! Finally, provided only pregnant mothers were supposed to answer the question and you are editing the response of a grandfather who, correctly, skipped the question, you may decide to treat this as 'not applicable'.

In other situations (or if you feel lazy), you may decide that you do not need to distinguish between any of these alternatives and just assign a single missing value for all four cases (indicating 'no answer' for whatever reason). Regardless of the outcome of your deliberations, it is important that missing values lie outside the expected range of legitimate answers (we shall come back to this point later in this chapter).

There are two other strategies for dealing with missing data. Depending on the computer package you plan to use for the analysis of your data, you may simply leave a blank and let the computer assign a so-called **system missing value**. Alternatively, you may decide that it is safe enough to plug in an actual value where no answer is given. For example, missing data on an income question is sometimes replaced with the average income of the other respondents. However, while this procedure has the advantage of not reducing the sample size as a result of item non-response, you have to be aware that, in essence, you are 'fabricating' your data! The problem is exacerbated when a large proportion of answers to a particular question is missing. To labour the point: when 80% of the respondents do not provide an answer to a question requesting their annual income, it is unacceptable to replace the missing data with the average of the other 20% who answered the question. So think *very* carefully before you embark upon a strategy of 'replacing' missing data with 'average' or 'typical' or 'likely' values!

The Role of Data Coding

Having carefully edited your data and corrected the many errors you have found, you should start thinking about **data coding**. This is something that most people leave to the last moment because, let's face it, coding questionnaires is not the most exhilarating task one can imagine (it only just beats sitting through a statistics lecture in entertainment value). However, coding is rather important, in that mistakes made here are difficult to correct at a later stage. In what follows, you will essentially learn two things: firstly, how to transform a bunch of completed (and edited) questionnaires into symbols which can be understood by a computer; secondly, how to find mistakes which are invariably made during data input.

data coding

To prepare for the transformation of answers into a computer-readable format, it is useful to imagine the final product first. A woman must have a vision (and, of course, a man, too)! After your splendid transformation efforts, you should end up with a data matrix like the one shown in Table 4.2. Note that the first line of bold figures is not part of the data matrix, it simply indicates the columns in which the data have been entered.

```
1234567890123456789012345 678901234567890123456789012345678 90 ...
001 1 1 2   700    1 2 2 9  5 Totally useless              4 ...
002 1 2 1  1300    4 3 1 4  5 Should be a must             5 ...
003 9 1 3  2100   11 5 1 1  3 Too expensive                1 ...
004 1 1 2  1800    2 4 4 4  4 I don't like academics       7 ...
005 1 1 2  9999    5 3 3 3  3 I never knew how to spell this  22 ...
006 2 1 1   600    2 2 3 3  2 My neighbour has one         5 ...
007 2 2 0   900   99 9 9 9  9 We should all walk          99 ...
```

Table 4.2   Example of a computer-readable data matrix

Looking at Table 4.2, it becomes apparent that a data matrix is rather useless unless we can establish a link to the answers in the questionnaire. For this, we strongly recommend that you set up a **code book** (sometimes also called a 'coding plan'). This gives 'a column-by-column explanation of the relation between the codes and the responses to the questions'.[2] Specifically, the code book describes how the responses relate to variables, labels i.e the variables, shows whether a particular variable is **numeric** (i.e a number) or an **alphanumeric** (i.e. a string of letters), gives the length of the variable specified (in terms of column width), and so forth. For example, the code book for the data matrix in Table 4.2 could look like the one shown in Table 4.3.

code book

numeric variable
alphanumeric variable

With the code book, the numbers in the data matrix (Table 4.2) can be interpreted. We now know that the first three numbers merely identify a particular case, in our example a questionnaire completed by a certain respondent. We can also see that the values in column 7 relate to the variable SEX and that 1 has been assigned to males and 2 to females (and not, as you might have expected, to 'regularly' and 'none'). But before you become too bushy-tailed and think you have cracked this one, let's have a closer look at the elements of a good code book:

Variable name.   The **variable name** is simply a name given to each of the

variable name

Table 4.3   Example of a code book

Variable name	Variable label	Value label	Type of variable	Column number	Missing value
CASENO	Case number	–	N	1–3	–
COUNTRY	Country of origin	Togo = 1 Haiti = 2 UK = 3 Syria = 4	N	5	9
SEX	Sex of respondent	Male = 1 Female = 2	N	7	9
CARS	Cars per family	–	N	9	99
INCOME	Monthly income (£)	–	N	11–14	9999
CHILD	No. of children	–	N	16–17	99
STATE1	Attitude to environment	Str. Agree = 1 Agree = 2 Neither = 3 Disagree = 4 Str. Disa. = 5	N	19	9
STATE2	Attitude to cars	As above	N	21	9
STATE3	Attitude to walking	As above	N	23	9
STATE4	Attitude to burial at sea	As above	N	25	9
COMMEN	Comments on catalytic converter	–	A	27–57	Blank
AGECAR	Age of car	–	N	59–60	99

WARNING 4.3

Do not define your variable names 'the quick way'. Any time you save on variable definition will be lost when you have to constantly (and frantically!) refer to your code book to remember which variable(s) you are dealing with.

variables in the data set so that the computer can reference them during the analysis. Note that some computer packages restrict the length of the variable name to 6 or 8 characters. So if you would like to call a variable describing the respondent's preferred winter holiday destination as MOSTPREFERREDWINTERHOLIDAYDESTINATION you may have to shorten it a bit (i.e. call it something like WINTDEST). Also, some programs do not permit you to use certain reserved words, usually those which have a particular meaning in 'computerspeak', such as AND, ALL, NOT, GE, LT or WITH. Further, you need to make sure that variable names are *unique* (i.e. appear only once); you should not give the same variable name to two (or more) variables as you are likely to create major havoc (computers are only human, they can be easily confused!).

Novices to data analysis are often tempted to take naughty short cuts in variable definitions and simply consecutively number variables (for example call them VAR1 to VAR723). However, unless you have an incredible memory for numbers or are one of those people who find great pleasure in learning rows of random numbers only to recite them at the annual convention of amateur pigeon breeders, we recommend that you refrain from this distasteful practice. During your analysis, you may not always see the complete variable label (see below) and to remember just exactly what VAR437 stands for appears rather more difficult than working out that AGECAR stands for 'age of car'.

Variable label. If your computer package permits you to use **variable labels** (i.e. a description of your variable), there is usually also a length restriction. However, this tends to be much more generous than with variable names. For example, in one popular analysis package called *SPSS* (see Chapter 5), the length restriction is currently 60 characters. Variable labels are useful in that if you forget what a variable name stands for, a look into your code book will tell you what it is (e.g. in Table 4.3, CHILD stands for 'number of children' rather than, say, 'instances of childish behaviour').

variable labels

Value labels. **Value labels** are descriptions of the values for each variable. They are not always necessary, particularly when it is quite obvious what the various values mean (e.g. as when you record the respondent's income in pounds or dollars). However, value labels need to be carefully defined for multiple-choice questions or scales in which you expect one particular answer out of a given number of alternatives. For example, going back to researching consumer perceptions to our new chocolate spread (Table 4.1), you will remember that we had the alternatives 'ecstatic', 'outstanding', 'exuberant', etc. We could now code 'ecstatic' as 1, 'outstanding' as 2, 'exuberant' as 3, etc. We could also code 'ecstatic' as 5, 'outstanding' as 4 or, indeed, as 4221. The particular number is irrelevant as long as we use different ones for the five alternatives (we are dealing with a nominal variable, remember?). By the same token, it does not matter whether you use 1 for 'male' and 2 for 'female' as we did in Table 4.3, or vice versa. However, for some types of scale, the way you assign values *is* important. Consider the two statements in Table 4.4 which are both measured on a Likert scale (see Chapter 3 for this scaling format).

value labels

HINT 4.3

When defining variable names, use easily recalled memonics, i.e. names that remind you of the meaning of your variables

Table 4.4 Attitudes towards driving and walking

	Strongly agree	Agree	Neither agree nor disagree	Disagree	Strongly disagree
STATE2 I like to use my car even for very short distances					
STATE3 Whenever possible, I leave my car and walk					

If we were to assign 5 for 'strongly agree' down to 1 for 'strongly disagree', a score of 5 would have very different meaning in the two statements. Specifically, it would indicate a liking for using cars in statement 2 and a dislike for using cars in statement 3. Thus, if we were to code both statements in an identical fashion (as we did in our code book), it would

HINT 4.4

Code variables relating to the same issue consistently, i.e. low values should represent unfavourable/negative views and high values favourable/positive views.

make coding easier but we would have to remember the *polarity* of each statement (for example, when comparing the average scores of, say, males and females). Similarly, if we wanted to aggregate the scales (i.e. sum up the values for each respondent), we would be forced to recode one of the variables so that, for example, low values always expressed 'anti-car' attitudes and high values 'pro-car' attitudes. Consequently, we recommend to code all questions relating to the same issue in such a way that low values express a low intensity (e.g. dislike of cars) and high values a high intensity (e.g. liking of cars).

Finally, it is important to have value labels for every possible type of response, including labels for missing data (see below). For data coded from open-ended questions, this often involves the specification of a category called 'other', to cope with responses that did not quite fit into the other categories you created.

Type of variable. We have already pointed out that there are two types of variables, namely numeric and alphanumeric (the latter also known as 'string' or 'alphabetic' variables). In Table 4.3, an example of the former type of variable is SEX, as we have coded 'male = 1' and 'female = 2', instead of, say, 'male = M' and 'female = F'. On the other hand, the comments regarding the catalytic converter (COMMEN), have been coded as an alphanumeric variable, i.e. typed in as they appear in the open-ended question (see Table 4.3). An alternative would have been to create specific response categories for these comments and generate a numeric variable. For example, we could have grouped them into 'positive', 'neutral', 'negative' and 'totally incomprehensible' and assigned a numeric value to each category (e.g. positive = 1, neutral = 0, etc.).

Which alternative is to be preferred depends on (a) the scope for sensibly categorizing open-ended responses and (b) whether you would rather have a listing of the original comments or work with fewer response categories in your analysis. There is a whole literature on how best to code replies to open-ended questions and several research methodology texts include extensive examples of coding instructions (see suggestions for Further Reading). For our purposes it is sufficient to point out that open-ended questions, while providing useful information, do not lend themselves easily to quantitative data analysis. The number of open-ended questions should therefore be minimized. Where they appear unavoidable, the responses to such questions should be classified into exhaustive and mutually exclusive categories; the latter should then be given numerical values to facilitate subsequent computer analysis.

HINT 4.5

Code your answers numerically whenever possible.

Column number. The column number(s) indicates the *location* and *length* of a particular variable in the data matrix. While this is relatively straightforward, a number of points ought to be made. Firstly, depending upon the computer package involved, it is not always necessary to leave spaces (i.e. empty columns) between consecutive variables. The reason we did this in our example in Table 4.2 is only to make our data look 'nice and tidy' (deep down we have a good core!) and to make it easier to detect mistakes (see later in this chapter). Secondly, some computer packages still require an 80-column format. Although most data input is nowadays conducted with screen editors, the 80-column convention (which is the maximum number of columns the good old punch card could take) frequently restricts the maximum length of a line of data. Of course, this raises the question what to do if one needs more space, for example because the researcher decided to observe 378 variables relating to the mating habits of the Dover sole. Basically, in such instances, the data can continue over a number of lines, but the computer package needs to be told about the total number of (80 column) lines per case. Exactly how this is done differs from one package to another; it is therefore necessary to consult the appropriate manual. Where cases require more than 80 columns, it is also advisable to repeat the case number at the beginning of each data line followed by a 'card' number for each line of data in a given case. If all this sounds too strange to you do not despair! Get yourself access to a decent analysis program which does not have these silly 80-column restrictions and accepts data input from a normal spreadsheet and/or interfaces with a tailor-made data entry module; there are several such programs as we will see in Chapter 5.

Missing value. The last entry in the code book specifies the values which represent missing data. In Table 4.3, in an attempt to keep our code book example simple, we have not distinguished between the four different types of missing data discussed earlier. To put this right, Table 4.5 shows an example of coding instructions considering different reasons for missing data.

Table 4.5 Coding an attitude statement

	Strongly agree	Agree	Neither agree/nor disagree	Disagree	Strongly disagree
STATE4 When my goldfish dies I would like him to have a respectable burial at sea			codes		
	1	2	3	4	5

Missing values:

forgot to answer	= 6	or (as in our code
don't know	= 7	book) one Missing
refused	= 8	value (9) for all
not applicable	= 9	four alternatives

As already mentioned, you should select your missing values labels outside the expected range of answers (see Table 4.5). Some people use negative numbers to make certain that their missing data codes are 'out of range'. We do not recommend this practice, as it requires more space (you need at least two digits for a variable) and introduces an additional source of potential input error. Note that we have used the number 9 as a missing value code throughout our code book example and matched the length of

HINT 4.6

Whenever possible use a consistent pattern for coding missing data.

each variable with a corresponding string of 9s (see Table 4.3, last column). While the specific number chosen is not particularly important, it permits us to develop a pattern by which 9s are immediately recognized as missing data.

Three final points before we leave the titillating topic of coding. Firstly, it is often advisable to build a little **coding template** (such as the one in Table 4.5) reminding yourself of the key coding instructions for each variable. This is particularly useful when you have several response formats and a large number of variables.

coding template

Secondly, you may consider not inputting your data directly from the completed questionnaires but transferring them initially onto **coding forms** (pre-printed forms specifically designed for data input). While this appears to be unnecessarily time consuming (since data entry is done in two steps), use of these forms tends to reduce mistakes relating to wrong column entries, particularly when large data sets are involved.

coding forms

Thirdly, still relating to large data sets, it may be advisable to investigate (*before* data collection) whether you can get access to a machine-readable input device such as an **optical scanner**. The beauty of this miracle of modern technology is that it can read data straight from your questionnaires (so you don't have to do any typing). However, if you go down this route you will most probably need to use rather special (i.e. computer-readable) forms to gather your data; this will push your data collection costs up considerably. However, this initial extra expense often pays off in terms of quicker, largely error-free data entry.

optical scanner

Finding Your Mistakes

Having coded your data and typed them all into the computer, you are likely to experience a big disappointment – you find mistakes in your data. Even if you are very careful during your data input, this is virtually unavoidable; the lengthier and/or more complex your questionnaire, the more likely it is that mistakes will be made during data entry. But do not give up, there are some useful techniques which you can apply to find certain mistakes.

Let us first focus on those mistakes which are relatively easy to find to provide you with an instant feeling of success. Such mistakes include skipping a number or an entire row; typing the same number or row twice; or putting numbers in what should be empty columns (i.e. columns being used to separate variables as, for example, in Table 4.2). To find mistakes of this kind, you should first check to see whether there is a discrepancy between the total number of cases in your data matrix (i.e. number of rows) and your sample size. You can do this by looking at a variable such as CASENO in Table 4.3 or, indeed, at any variable (the total number of legitimate answers plus the total number of missing values should always equal your sample size; if not, you have either omitted or duplicated one or more questionnaires). Next, you should define a few (temporary) **test variables** for those columns which you expect to be empty and ask your computer to calculate, say, an arithmetic mean, for these; if you get *any* number, you know you have made a mistake (because no data should have

test variables

been entered in these columns). You should also calculate the minimum and maximum values for all your variables and compare them against your code book to see whether any of them are 'out of range' (e.g. if you are dealing with a 5-point scale anchored at 1 and 5, respectively, and your computer tells you that its minimum is 0 and its maximum 7, you have inputted wrong values somewhere). Finally, if you want to be *really* thorough, you should try to set up a few 'IF . . . THEN . . .' routines. For example, going back to our code book in Table 4.3, you could check that 'Age of Car' is coded as 99 (missing value) when the value of 'Cars per Family' is 0.

Mistakes which are not easy to find are those where values have been entered which are within the expected range but, nevertheless, wrong (e.g. entering a 3 for a response relating to a 5-point Likert scale while the correct response was 4). With such errors, only a very time-consuming one-by-one comparison of the questionnaires (and/or coding sheets) with the data matrix in your computer can help. Nevertheless, you should at least select a few of your questionnaires at random and compare them with the corresponding entries in the data matrix; the more errors you detect, the less proud you should be about your coding and data entry performance.

WARNING 4.4

Data coding and entry is bound to cause mistakes. Time spent detecting and correcting errors is time well spent – so do not take chances by leaving your data matrix unchecked.

Transforming Variables

Having gone through the painful processes of editing, coding and finding input errors, you now deserve to have some fun. Changing some variables and/or creating new ones is the perfect excuse to let your creative side loose! **Variable transformations** are often necessary to carry out a particular analysis and/or convenient to facilitate data reporting. To illustrate their role, let us assume that you have gathered data on the different sources of income for a sample of business executives. At present, you have the five separate variables listed in Table 4.6, each of which records income in pounds.

variable transformations

Variable name	Variable label
BLMAIL	Blackmail
EXTOR	Extortion
PROSTI	Prostitution
BRIBE	Bribery
LECTUR	Occasional lectures on business ethics at various universities

Table 4.6 Different sources of income

If you plan to analyse the overall income of this profession (in order, say, to compare it with the average annual income of a bagpiper in the 3rd Royal Scottish Highland Battalion), you need to create a new variable (call it TOTINC); this would be defined as the sum of the five original variables, i.e. TOTINC = BLMAIL+EXTOR+PROSTI+BRIBE+LECTUR. Note that

TOTINC is an *additional* variable, with the five original variables still existing unchanged.

If we were to focus particularly on the income generated through bribery (BRIBE) and decided to show the income distribution with the help of a chart, we would need to form appropriate categories (as income is a continuous variable). For example, we could recode BRIBE as shown in Table 4.7. Note that if such recoding is permanent (i.e. we save the recoded variable under the same variable name), the character of the original variable has been altered irreversibly. In our particular example, a ratio scale variable has now been downgraded to an ordinal variable with only four values. Thus, the more detailed information collected originally (i.e. the exact income of each respondent through bribery) would no longer be available. As already mentioned in Chapter 3, apart from the actual information loss, such recoding would also severely restrict the scope for future analysis (e.g. with the new ordinal-scaled variable one could no longer calculate the average income achieved through bribery).

Table 4.7   Recording a variable

Bribe (£)	New categories
0–999	(1) Negligible bribery income
1 000–4 999	(2) Low bribery income
5 000–9 999	(3) Medium bribery income
10 000–maximum	(4) High bribery income

Finally, you should explore the scope for adding variables to your data set from secondary data. In industrial marketing research, for example, company information like number of employees, sales revenue, principal activity, etc. is often available from published sources (e.g. business directories) and there is no need to waste precious questionnaire space and/or risk alienating your respondent by requesting this information again. There is nothing stopping you from augmenting your data matrix by adding variables relating to your respondents from whatever source and you should try to identify appropriate sources. Of course, you need to ensure that the data obtained from such sources are comparable to your own. For example, relying on a directory published in 1973 to obtain much-needed sales revenue information relating to firms surveyed during 1994 is not exactly a brilliant idea.

Summary

Having demonstrated that there is more to editing, coding and data input than meets the eye, how can we best summarize the key points? (This is a purely rhetorical question – after all, we are paid to write this book and should know the key points!) First, we focused on data cleaning or editing, which attempts to detect and correct data errors. A distinction was made between field edits and central office edits and various editing tasks were discussed (including dealing with ambiguous data, logically inconsistent data and item non-response). We then moved into data coding and showed you the value of setting up a code book, and followed this by a discussion of strategies for finding mistakes resulting from data entry. Finally, we discussed different variable transformations and potential pitfalls associated with them. Throughout this chapter we made implicit reference to the use of computers in data analysis but did not explain *how* they can be of benefit to you. This is what Chapter 5 is all about.

Questions and Problems

1. What are the advantages of a field edit over a central office edit?

2. Distinguish between ambiguous answers and logically inconsistent answers and give two examples of each. What steps can you take to avoid such problems?

3. What are the four different forms of item non-response?

4. How would you code missing data when entering questionnaire results into a computer?

5. Construct an example to illustrate the difference between a variable name, a variable label and value label.

6. Under which circumstances would you change the polarity of attitude statements during the coding process?

7. What are alphanumeric (string) variables and how would you go about coding them?

8. What are typical mistakes that can occur during coding? How would you go about identifying them?

9. Under what circumstances would you consider transforming your variables? Give three examples of variable transformations.

10. What advice would you give to fellow sufferers to relieve boredom during coding?

Notes

1. Aaker, DA, Kumar, V & Day, GS (1995) *Marketing Research*, 5th edn, p. 443. New York: Wiley.

2. Lehmann, DR (1989) *Market Research and Analysis*, 3rd edn, p. 369. Homewood, IL: Irwin.

Further Reading

Bourque, LB & Clark, VA (1992) *Processing Data: The Survey Example*. Newbury Park, CA: Sage Publications. A comprehensive guide on how to handle problems associated with coding.

Knaus, R (1987) Methods and problems in coding natural language survey data. *Journal of Official Statistics*. Provides a good discussion on how to code open-ended questions.

Steward, DW (1982) Filling the gap: a review of the missing data problem. In BJ Walker et al. (eds), *An Assessment of Marketing Thought and Practice*. Chicago: American Marketing Association. Everything you ever wanted to know about handling missing data.

5

Do you have access to a *computer package* or must you employ Bill Gates?

The Role of Computers in Data Analysis

Most of you would agree that you do not need to understand the intricate details of modern car engine technology to be a competent driver who enjoys driving. It would also not occur to you to take an expert car mechanic along with you every time you go for a drive (unless, of course, you have a crush on car mechanics and are guided by ulterior motives). But once in a while, something will fail and you will have to seek the help of a mechanic.

With the advent of modern computer technology and the proliferation of user-friendly hardware and software, the same applies to data analysis. The overwhelming majority of time, you will find that you do not need any help from experts, who usually come in the strange shape of computer specialists and statisticians. You will be able to handle your data input, analyses and reporting very competently yourself, thank you very much. In fact, most of the time you will presumably be better off to do the work yourself through developing an understanding for the data at hand and avoid being bored stiff by 'experts' who often have an uncontrollable urge to tell you everything *they* know about a particular subject rather than what *you* need/want to know. However, just as with cars, you also must recognize your limitations and, once in a while, you will have to resort to experts. This chapter will help you identify those tasks you can handle yourself and those with which you may need to ask for help. It should also put you in a position to ask more intelligent questions and, hopefully, get more meaningful answers.

Broadly speaking, there are four types of packaged computer applications you should find useful:

1. Applications which will help you to *input your data* in a convenient way and in a format suitable for subsequent analysis (see also the highly exciting discussion on coding in Chapter 4).
2. Programs which will enable you to actually *carry out the analysis*. Even experts with a strong quantitative background routinely use such packages to save time when dealing with anything but the simplest statistical analyses, so why shouldn't you?!
3. Packages which will help you to *report your findings* in an appealing way, either in written/tabulated form or with the help of fancy

graphics. Presenting your results in an attractive form can only impress the reader and works always to your advantage (we will have much more to say about this in Chapter 15).

4. Flight simulators, pigeon shooting and other computer games which have absolutely nothing to do with data input, analysis or reporting but have great entertainment value and help you to relax. For obvious reasons, we must sadly confine ourselves to a discussion of the first three groups of applications only.

Most of the above packages are available for personal computers (PCs) and, unless you have a monstrous amount of data to analyse or cannot get access to a PC, there is no need to use a large, frequently less user-friendly mainframe system.

Data Input Packages

spreadsheets

cell

Just before we get ourselves into trouble (again), we need to point out that the types of packages we are talking about here often fulfil other functions in addition to data entry. **Spreadsheets**, for example, are often used for data input; however, they are primarily designed to manipulate data matrices, e.g. add data columns, multiply certain numbers with others, transform data into graphs and even perform some (basic) statistical analysis on the data (furthermore, you can use them for some incredible hide and seek games to amuse your children or bamboozle your grandmother). Yet, for our purposes, we are merely concerned with their ability to facilitate data input. The usefulness of spreadsheets becomes apparent when one visualizes their typical format, which consists of a grid of rows and columns, each specifying a specific column–row address called a **cell**. Figure 5.1 illustrates a typical spreadsheet format in which cell D2 is highlighted.

Figure 5.1 Typical spreadsheet format.

Specifying a column for each variable and a row for each case permits you to type in your data with great ease; in a sense you are given the 'skeleton' of the data matrix (see Table 1.2 in Chapter 1) and all you have to do is type in the data values. Most spreadsheets nowadays are relatively large, i.e. they have a sufficient number of rows and columns to handle quite large data sets. Obviously, this is something which you should find out *before* you start inputting your data, otherwise you may run out of space (and this can be very frustrating if you have already input 99% of your data but can't squeeze in the last 1%). The size of a particular spreadsheet is usually given in the appropriate user manual (primarily because they now approach the size of a football pitch and it always impresses people). If you do not have a user manual at hand or find it too taxing/time- consuming/boring to read one, just scroll your cursor to the last cell on the bottom-right of your grid; if it takes you five minutes to reach it, then your spreadsheet is probably large enough.

Some of the better-known spreadsheets are *Lotus 1-2-3*, *Excel*, and *Quattro Pro*. However, there are many more and, since we are looking at them merely as data input devices, it does not really matter which particular one you use as long as (i) it has the capacity to handle your data and (ii) it interfaces with the analysis program you are going to use (see next section).

An even more convenient way to enter data is through the use of software packages that are purposely designed to interface with a data analysis program. An example of a **data entry package** is *SPSS Data Entry*, which includes user-friendly facilities for data input and editing on your PC and interfaces with the *SPSS/PC+* statistical analysis package (see below). In addition to data input itself, this package also makes definition of your variables disgustingly easy (see Chapter 4 on defining variables names, labels, etc.). Figure 5.2 shows how you are prompted by *SPSS Data Entry* when you want to define a variable.

WARNING 5.1

Make sure that your spreadsheet is large enough for your requirements *and* that the particular spreadsheet version can be read by the data analysis package you intend to use.

data entry package

Current File

STARSTUDY.SYS

Variable Name	
Variable Label	
Type of Variable	Numeric
Variable Length	1
Decimal Places	0
Display Mode	Edit
Missing	

Press ^F10 to complete

Figure 5.2 *Variable definition using SPSS Data Entry.*

After opening the file in which you are going to store your data (in the example of Figure 5.2, this file is called STARSTUDY.SYS), you are guided through all stages required to specify a variable. For some headings, defaults are already provided (e.g. it is assumed that the variables are numeric and not strings) but they can, of course, be overwritten.

Inputting the actual data is equally convenient. Specifically, a spreadsheet-type input format is automatically generated, with each column referencing the variables previously specified; all you need to do is fill in the cells (no, the program *cannot* do this for you!). As an example, Figure 5.3

shows the computer screen for inputting the values for the first few variables included in our code book example (see Table 4.3 in Chapter 4). Note that the display extends automatically (i.e. scrolls) past the boundaries of the screen when you move the cursor to a data cell that was not visible before. Once the process of entering values is completed, the fully labelled file can be directly read by the *SPSS/PC+* data analysis package. Most 3-year-olds should be able to handle this without too many problems!

Figure 5.3 Data input using *SPSS Data Entry*.

STARSTUDY.SYS

CASENO	COUNTRY	SEX	CARS	INCOME	CHILD	STATE1	STATE2	STATE
001	2	1	3	1500	3	5	3	2
002	2	2	2	1800	2	2	1	3
003	1	1	1	9999	0	1	3	4
004	1	1	2	1200	4	2	2	4
005	1	2	0		.	.	.	.

Press ^F10 to complete

text editor
word processor

In addition to using spreadsheets or dedicated data input packages, there are other ways to create data files. You could, for example, use a **text editor** or a **word processor**. If you are familiar with a particular package, you can use this to punch in your data. However, make sure that you save your data as ASCII or a text file with no specific format codes (e.g. highlights, italics, underlines), otherwise you may have problems in transferring the data to an analysis package. Again, it is important to ensure *before* you use a word processor that the data analysis package you intend to employ can subsequently read the data in this format (see also Warning 5.2). Ask an expert if you are uncertain or if you are looking for a good excuse to chat-up your favourite classmate. Overall, using general text editors and word processing packages for data input is clearly a second-best alternative when compared to tailor-made data entry programs like *SPSS Data Entry*.

Data Analysis Packages

Computer packages handling the actual statistical analysis come in two basic forms: specialized programs which focus on a particular application, and general 'workhorses' which offer a wide variety of facilities. Examples of specialized packages are *Forecast Pro* which focuses, not surprisingly, on forecasting, *BestFit*, specializing on fitting curves to data, and *Supertree*, a decision tree analysis package (there's plenty more around and the Further Reading section will tell you where to find them).

While most of these specialized packages are undoubtedly very impressive, they are only designed to do a few things and, therefore, are only recommended for users with very narrowly defined data analysis needs. The large majority of aspiring analysts (such as yourself) will be infinitely better off using widely available 'general purpose' packages, which can certainly do all the different types of analyses discussed in this book and many more. Such packages have facilities for data tabulation, descriptive statistics, cross-tabulations, tests for differences between groups, data plotting procedures, correlation analysis, plus a variety of non-parametric tests and multivariate statistics; in short, more than the ordinary mortal would use in a lifetime. Among the best-known programs are *MINITAB*, *SAS*, *SPSS* and *Statgraphics*. There are different versions of most of these for mainframes and PCs and for different operating systems such as *DOS* and *Windows* (for example, the *DOS* version of *SPSS* is known as *SPSS/PC+*, while its *Windows* counterpart uses the highly original name *SPSS for Windows*).

If you are working in a university or any large organization, chances are that your institution will already have a general purpose analysis package (if not, you should seriously consider whether you are working for the right guys!). In terms of user-friendliness, the programs are good *once you have defined your data*, i.e. coded and labelled all your variables, values, etc. For example, to produce a tabulation (or so-called frequency listing) of the variables COUNTRY and SEX after you specified your data and saved the lot in a file called STARSTUDY.SYS (see Figure 5.3), the *SPSS/PC+* package – which is conveniently menu-driven – only requires a two-line command, as shown in Table 5.1.

```
GET FILE = 'STARSTUDY.SYS'.
FREQUENCIES VARIABLES = COUNTRY, SEX.
```

Table 5.1 *SPSS/PC+* commands for a frequency listing

The output for this simple command is given in Table 5.2 (for the other general purpose packages, the commands are equally straightforward). Thus, the critical issue tends to be data definition and not the commands required to conduct a particular analysis. As already mentioned, you will save both time and effort if you can use a data entry program which interfaces with a particular analysis package and also enables the definition and labelling of variables all in one go (e.g. the *SPSS Data Entry* software mentioned earlier).

If you cannot get your hands on dedicated data entry software or if the data is already supplied to you in a spreadsheet format or as an ASCII file, make sure that your data analysis program will be able to read the data. With some data analysis packages, it can be difficult or even impossible to 'translate' differently stored data into the required format. Other packages, on the other hand, can read virtually everything (including some pretty weird formats). Finally, you should ensure that you stay within the processing limits of your analysis package. For most applications, this is not a problem. However, if you are dealing with very large data sets (because you collected heaps of unnecessary information or object against sampling on religious grounds), it is worth checking whether there is a limitation in the number of variables and/or cases which can be analysed by your particular package *at any one time* (you should be able to find the answer in the user manual – if not, call the software supplier and demand some service!).

HINT 5.1

Whenever possible, use a data entry package which also guides you through the definition of variables and interfaces easily with your analysis package.

WARNING 5.2

Find out whether your analysis program (a) can read the data in the format you have used/intend to use for input, and (b) has the capacity to analyse the number of cases/variables in your data set.

Table 5.2 *SPSS/PC+*
frequency tables for COUNTRY
and SEX

Country   Country of origin

Value label	Value	Frequency	Percent	Valid percent	Cumulative percent
Togo	1	393	26.7	26.8	26.8
Haiti	2	476	32.3	32.4	59.2
UK	3	253	17.2	17.3	76.5
Syria	4	345	23.4	23.5	100.0
Missing data	9	6	0.4	MISSING	
	Total	1473	100.0	100.0	

Valid cases 1467   Missing cases 6

Sex   Sex of respondent

Value label	Value	Frequency	Percent	Valid percent	Cumulative percent
Male	1	598	40.6	40.6	40.6
Female	2	875	59.4	59.4	100.0
	Total	1473	100.0	100.0	

Valid cases 1473   Missing cases 0

Presentation Packages

It is not enough to have generated pages and pages of more or less (usually less) sensible computer printout; you will have to communicate your findings as well (see also Chapter 15). This usually means that you have to compile a report, include some flashy graphs and, perhaps, even produce some overheads or slides. A number of software packages (as well as a supportive secretary and loads of coffee) can become handy at this stage. To begin with, you will definitely need a word processing package to type your report. It is most likely that you already have such a package installed on your PC (unless your PC is so old that it still takes solid fuel to operate it). Well-known word processing packages are *Wordperfect* or *Word*, but, whatever you use, it is best to stick with the particular package you happen to know (unless, of course, you get immense pleasure from spending hours on end trying to figure out how to use italics or set page lengths with an unfamiliar package). And a hint for the 98% who belong to the convenience-oriented segment of our readership: try to transfer the file(s) in which you have stored your analysis results (i.e. results files or output files) directly into your word processing package. Usually, this works without problems and it often saves retyping tables.

If you really want to impress with your report, you may consider using a desktop publishing application such as *Pagemaker*, *Ventura*, or *Interleaf*. As their name implies, these applications can produce highly sophisticated documents which are bound to cause gasps of amazement by all who see your final product! As a final touch, for producing nice overheads and graphs you may also like to familiarize yourself with one of a whole range

HINT 5.2

Try to transfer your results files (or parts of them) into your word processing package. This saves some retyping of results when you write up your analysis.

of specialized graphics/charts packages. *Harvard Graphics*, *Power Point* and *Freelance* are the most popular, but there are plenty of other nice programs around.

Note that, like computer games, presentation packages can become addictive: once you start 'playing' with presentation tools you may be tempted to spend hours improving your charts through adding more and more impressive features, rather than doing more (or more useful) analysis. However, at the end of the day, an excellent presentation of a poorly-conceived and/or executed analysis is not something to be proud of.

To help you in your deliberations with choosing a package to suit your needs, Table 5.3 lists several factors that you should consider (particularly if you are thinking of parting with some of your own hard-earned cash to buy a package or two). Moreover, since software becomes as quickly outdated as Posh Spice's wardrobe, you should try and keep up with new developments (this should also prevent unscrupulous salespeople from dumping hopelessly obsolete and/or overpriced packages on you). Good sources of information are the *Directory of Software for Marketing & Marketing Research,* which is regularly published by the *American Marketing*

Table 5.3 Choosing software for analysis

	Type of package		
	Spreadsheets	Statistical analysis	Graphics
Technical requirements			
1. Compatibility with operating system	X	X	X
2. Required memory	X	X	X
3. Printer compatibility	X	X	X
4. Need for a special graphics card			X
Performance			
1. Size of matrix (rows/columns)	X		
2. Scope for combining different matrices	X		
3. Scope for depicting different graphs			X
4. Restrictions in the number of data points per graph			X
5. Editing/labelling capabilities			X
6. Number of standard functions	X	X	X
7. Available statistical methods		X	
8. Data reporting options	X	X	
9. Printer/plotter options	X	X	X
10. Data transfer to and from other programs	X	X	X
11. User-friendliness of editor (full-screen)		X	
12. Window capabilities	X		
User-friendliness			
1. Help functions	X	X	X
2. Tutorials	X	X	X
3. Handbooks/user guides	X	X	X
4. Menu/mouse driven	X	X	X

Association (AMA), and the *OR/MS Resource Directory* published by the *Institute for Operations Research and the Management Sciences* (INFORMS); for details, see the Further Reading section.

Two final points apply to everything we have discussed in this chapter. First, several of the newest versions of data analysis software (e.g. *SPSS for Windows*) are of the 'all-in-one' variety, i.e. they are **integrated packages**, encompassing data definition and input, statistical analysis, as well as graphics. Such packages have the definite advantage that you only need *one* piece of software with which to run several applications and you can usually operate them by means of a mouse (the technical not the rodent variety!). The downside is that they tend to be rather hungry in terms of memory requirements and, unless your PC is pretty fast, they can take their time in running applications.

Second, no matter what software you use for whatever purpose, *make sure you have back-up copies of all your important files, be it the data file, the command files you used to run the analysis or the output/report files*. Disks occasionally get damaged in disk drives, are often mistaken for bath toys by young children, or get chewed-up by your pet alligator (so, do not risk carrying your entire PhD work on a single floppy). Hard disks are not invincible either, particularly if you happen to spill hot coffee over them or have bought your PC in a car boot sale.

integrated packages

WARNING 5.3

Always make *several* back-up copies of your important files and do not keep all copies in the same place.

Summary

In this chapter we have undoubtedly convinced you that modern computer technology has made data analysis relatively straightforward and easy. There are several possibilities for inputting your data, a number of very powerful data analysis packages and a near-endless supply of software which can help with reporting and presenting data. In terms of data input, programs which also guide the definition of variables, labels and missing values are strongly recommended, particularly when they interface directly with a good analysis package. For data analysis, most people will find general purpose programs rather than single-focus applications more useful. In terms of reporting, a good word processing package is a minimum requirement, while desktop publishing and/or specialized graphics programs are the business for producing professional-looking documents. If you want maximum flexibility and minimum hassle, go for an integrated package which should keep you busy for a while.

Questions and Problems

1. What types of computer packages can you use for data entry purposes?

2. Is a word processor preferable to a spreadsheet for inputting data?

3. Name three general purpose statistical packages that are widely used for data analysis.

4. Why is correct data definition so critical when using a statistical package?

5. Why is it important that your statistical package, word processing software and presentation software are compatible?

6. What features determine the user-friendliness of a software package? Which is the most important to you?

7. What features of your computer must you bear in mind when deciding which statistical package to use?

8. From where could you obtain a list of software packages for data analysis puposes?

9. If you have access to the Internet, try and use it to answer Question 8.

10. Have you backed-up your files?

Further Reading

American Marketing Association (1995) 1995 Directory of Software for Marketing & Marketing Research. *Marketing News* (10 April): 11–24. Annual shopping guide which lists loads of suppliers of software as well as a description of the products they offer.

Babbie, ER (1995) *Adventures in Social Research: Data Analysis Using SPSS for Windows*. London: Sage Publications. If you are going to use this package, this book is the business.

Institute for Operations Research and the Management Sciences (1995) 1995 OR/MS Resource Directory. *OR/MS Today* (October): 75–96. Listing of both specialized and general purpose software according to topic, together with supplier profiles. Note that *OR/MS Today* regularly publishes software reviews and comparisons of different packages, so check it out next time you visit your library.

6

Why do you need to know your *objective* before you fail to achieve it?

The Need for Analysis Objectives

So here you are, sitting in front of your computer, raring to go; your data, meticulously edited and coded sit restlessly in the depths of your computer's memory and eagerly await your instructions to reveal their secrets. Your problem at this stage is to decide exactly what instructions to give, in **analysis** other words, what **analysis** to undertake.

One easy (but certainly not recommended) solution to your problem is to follow the 'pack the kitchen sink in' approach, also known as the 'run anything and everything' strategy. This approach simply involves asking your computer to perform all known statistical tests in the universe on all your variables – including relating everything to everything else (the rationale being that by producing 128 miles of computer printout something useful is *bound* to turn up!).

Unfortunately, this approach has several drawbacks, the most important of which is that it doesn't work! The reason for this is that

WARNING 6.1

Don't be tempted to use your computer's capability to run 'everything under the sun'. You will live to regret it.

'The computer will do only what it is instructed to do, and the validity of the results it generates depends on the quality of the information fed into it and the skill with which it is instructed to analyse the data. If faulty methods of analysis are employed, the results may not improve decisions no matter how rapidly the computations are made'.[1]

analysis objectives So, what's the alternative? Well, the alternative is to set some **analysis objectives** – preferably some good ones!

If you think of analysis as the process by which one makes the data 'talk', setting analysis objectives is deciding exactly what kind of information one wants to get out of the data. Data can be made to 'talk' in different ways, depending upon how they are manipulated. The choice of the particular technique depends, above all, on the specific needs of the researcher/ analyst, i.e. the sort of answers that he/she seeks from the data (to be sure, the choice of analytical technique depends on other factors as well, as we will see later on).

Clear analysis objectives serve to direct and guide the analysis process and are absolutely essential for the success of the latter. Specifically, analysis objectives fulfil three roles. First, they help ensure that only relevant analysis is undertaken; relevance refers to the extent to which any analysis performed contributes directly to answering the research questions of interest. For example, if you are seeking to determine what criteria company executives use in their choice of massage parlour, a cross-tabulation analysis of age by sex does not make you any the wiser about their choice criteria.

Second, analysis objectives provide a check on the comprehensiveness of the analysis. Comprehensiveness refers to the extent to which the set of analyses performed makes full use of the information potential in the data. Sticking with the same example, if you had data on both the sex and the age of the respondent but only examined the impact of sex (or age, but not both) on the choice criteria, then your analysis would have been relevant but not comprehensive.

Third, analysis objectives help avoid redundancy in the analysis; redundancy refers to the extent to which different analyses overlap, that is provide essentially the same information. In our, by now famous, massage parlour example, if you first compared the number of male executives considering criterion X (e.g. 'discrete location') as important with that of females and then contrasted the number of males and females *not* considering criterion X as important, then your additional contribution to knowledge would be precisely zero (because the content of the second analysis is already fully contained in the first).

In short, having analysis objectives helps ensure (a) that you only do relevant things, (b) that you do enough, and (c) that you don't do things twice.

Setting Analysis Objectives

The starting point for setting sound analysis objectives should always be the overall research objectives. Most of us do not wake up one Monday morning with an uncontrollable desire to analyse a data set we have never seen before and know nothing about. At some point in the past we somehow got involved in a research project and got hold of some data, and now we actually have to do something with it (which is a hassle, but that's life for you). Obviously, there must have been a purpose to the research project (although, by now, you may have forgotten what it was!). The **research purpose** indicates why the study was carried out in the first place and what it hoped to achieve.

research purpose

A careful re-examination of the overall aims of the research provides an excellent point of departure for developing analysis objectives. In this context, there has to be an explicit link between the analysis objectives and the overall research objectives, in the sense that achievement of the former should contribute towards achievement of the latter. Clearly, the better-specified the original research objectives, the easier it is to derive appropriate analysis objectives from them. For example, a study aimed at 'investigating the export behaviour of nightgown manufacturers in Afghanistan' provides hardly any clue about exactly what it hopes to achieve (other than that it has to do with exporting). Although more insight may be gained by looking at the specific variables that have been included in the study, one is still left unclear as to (a) why were these specific variables considered (and not others), and (b) what is one supposed to do with them (i.e. what kind of information should one be looking for). The problem here is that the research purpose has been too broadly defined and not translated into concrete research objectives; as a result, practically any set of analysis objectives would be consistent with the study's aims.

If, on the other had, the same study had been defined as aiming to 'compare Afghan-owned and foreign-owned nightgown manufacturers in terms of (a) company size, (b) export experience, (c) extent of export operations, (d) resources allocated to exporting, and (e) export performance', then the development of appropriate analysis objectives would be far less problematic. To begin with, one would immediately know that the main focus should be on *comparisons* rather than an overall description of the nightgown manufacturing industry in Afghanistan. Moreover, one would know what to look for in the variable list (i.e. specific indicators of size, export experience, export operations, and so on). Finally, one would be able to identify the limitations of the analysis (e.g. inadequate representation of 'export operations' due to, say, omission of the 'number of export destinations' from the list of variables).

WARNING 6.2

If you get involved in an ill-defined research project, chances are that you will encounter problems at the analysis stage. So, don't!

It is helpful to think of the process of setting analysis objectives as involving two sorts of decisions. The first relates to the **content of** an intended **analysis** and involves the selection of the variable (or set of variables) to be included in it; clearly, an analysis objective would be of little help if it did not reference the particular variable(s) to be analysed. The second decision relates to the **focus of** the intended **analysis** and involves the specification of the analytical stance or orientation to be adopted (e.g. simple description versus examination of relationships). Together, content and focus enable *operational* analysis objectives to be set such as 'describe the sample/population in terms of variables X, Y and Z'; 'break the total sample into two sub-samples according to variable A and compare them in terms of variables B and C'; 'examine the relationship between variable D and E'; and so on. Of course, you should always make sure that such specific objectives *do* indeed satisfy the criteria of relevance, comprehensiveness and avoidance of redundancy mentioned earlier.

content of analysis

focus of analysis

The Question of Focus

While the content of any analysis is, by definition, project-specific (since it depends on the particular variables that have been included in the study), its focus can only take one of three basic forms: description, estimation and hypothesis-testing.

With a **descriptive focus**, as the name implies, the aim is to paint a summary picture of the sample (or population) in terms of the variable(s) of interest. Say, for example, that you have randomly selected 200 university students and conducted interviews on their disco-going habits. When you come to analyse this data, one thing you may want to do is to provide a sample breakdown in terms of, say, sex, faculty, year of study and disco-going frequency. Statements that 120 members of your sample are female; that two-thirds of all respondents are matriculated in the Faculty of Theology; that the proportions of first, second and third-year students are

descriptive focus

20%, 30% and 50% respectively; and that, across the entire sample, the average disco-going frequency is 4.6 times a week (during term time, of course), are all examples of a descriptive focus. There are a number of statistical techniques one can be used to undertake descriptive analysis; among the connoisseurs in the field these are known as **descriptive statistics** and will be discussed in Chapters 7 and 8.

With an **estimation focus**, again as the name implies, the aim is to use the information one has on the *sample* to estimate the situation that is likely to exist in the *population* as a whole. For example, given that the average disco-going frequency of 200 randomly selected students turns out to be 4.6 times a week, what is likely to be the average disco-going frequency across *all* students in the university concerned? Thus, estimation can be seen as the process of making an informed guess based on incomplete information. The guess is informed because we use the information we have on the sample (together with some statistical theory to be introduced in Chapter 8) to say something about the population from which it was drawn. At the same time, the very fact that we use a sample means that our information is incomplete because it is not based on the entire population. In other words, our information contains sampling error and, therefore, it does not perfectly reflect the situation in the population (we talked about this in Chapter 2, remember?). However, to the extent that the sample has been drawn probabilistically (as is the case in the above example), then we can *assess* the likely sampling error and incorporate it in our calculations of the population estimates, ending up with **confidence intervals** for the latter. How this is done is the subject of Chapter 9, so don't be impatient!

Finally, with a **hypothesis-testing focus**, the aim is to test specific propositions concerning the variables of interest and use the evidence provided by the sample to draw conclusions regarding these propositions for the population as a whole. For example, you may hypothesize that male and female students differ in terms of their disco-going frequency (i.e. that there is a relationship between gender and going to the disco). If, say, you found out that the 120 female students in your sample went to the disco 4.8 times a week on average, while the 80 males went only 4.3 times, could you conclude that female students *in general* tend to visit the disco more often than males? Well, your sample results suggest that this might be the case, but what about sampling error? How likely is it that the difference in disco-going frequency observed in the sample actually reflects a 'true' difference in the population, rather than merely sampling error? To answer this and similar questions, we apply what are known as **significance tests** which are statistical techniques designed to help us decide whether our sample results are likely to hold in the population as a whole. Again, as was the case with estimation procedures, it is assumed that our data are based on a probabilistic sample so that assessment of sampling error is feasible. We will come back to the issue of hypothesis testing in Chapter 10.

When the focus of analysis is on estimation or hypothesis-testing, we use our sample to make *inferences* about the population. This process is formally known as **statistical inference** and the various techniques that are employed are commonly referred to as **inferential statistics**. Without

descriptive statistics

estimation focus

confidence intervals

hypothesis-testing focus

significance tests

statistical inference
inferential statistics

inferential statistics, the only other way for making statements about the population is to conduct a census, i.e. obtain data on each and every population element; for reasons already discussed in Chapter 2, this is usually not feasible and, therefore, inferential statistics are indispensable in data analysis.

Choosing the Method of Analysis

Although well-specified analysis objectives (in terms of content and focus) are essential prerequisites for successful analysis, there are additional factors to consider before an appropriate analytical technique can be selected. Figure 6.1 provides an overview of the factors bearing on the choice of analytical technique; while their specific influence will become clearer as we go through Chapters 7 to 13, we felt that you would be eternally grateful if you could have a first taste in this section.

Figure 6.1 Factors influencing the choice of analytical technique.

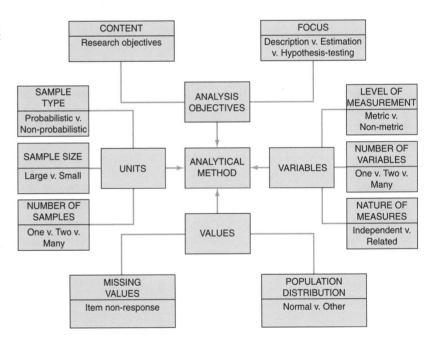

To begin with, the characteristics of the sample in terms of type and size will affect the choice of technique. With regard to *sample type*, as you should know by now, unless the sample has been drawn probabilistically the use of inferential statistics is not legitimate, since the latter make use of the concept of sampling error which – as pointed out in Chapter 2 – cannot be assessed where non-probability sampling methods are employed. What this means is that you will be treading on very thin ice if you try to estimate population parameters or test hypotheses with a non-random sample (e.g. a convenience or judgmental sample).

With regard to sample size, suffice it to say at the moment that some statistical procedures do not work well, unless one has a 'sufficiently large' sample. In practice, for simple analyses, this translates to a sample size of at least 30. The procedures susceptible to small-sample problems are a set of inferential procedures known as **parametric statistics** which are used for estimation and hypothesis-testing purposes. Thus, with small samples one may be limited to using only **non-parametric statistics** (sometimes also known as 'small-sample' statistics) which are less powerful techniques (we briefly touched upon parametric and non-parametric statistics in Chapter 3, when we distinguished between metric and non-metric measures).

A final sample characteristic affecting the choice of analysis technique is the *number of sub-samples* that one may wish to consider in a particular analysis. This is particularly relevant when the aim is to undertake comparisons among different groups in one's data set (e.g. male v. female, buyers v. non-buyers, English v. Welsh v. Scots, etc.). As will be discussed in Chapter 12, different statistical procedures are appropriate when only two groups are to be compared as opposed to when three or more groups enter the comparison.

Another influence on the method of analysis comes from the *measurement characteristics* of the variables involved. As was pointed out in rather excruciating detail in Chapter 3, the higher the level of measurement, the more sophisticated the analysis that can be applied to the data; and, yes, parametric procedures only work with metric (i.e. interval and ratio) data (as another look at Hint 3.2 in Chapter 3 should quickly confirm).

A related consideration is the *number of variables* that one wants to simultaneously analyse *and* the extent to which these differ in terms of their level of measurement. When one is interested in a single variable at a time (what we call **univariate analysis**), then there is no real problem because there is only one level of measurement to consider. Things get a bit more complicated, however, when two variables are involved (the case of **bivariate analysis**). As can be seen from Table 6.1, six distinct combinations of levels of measurement are possible, each requiring a different statistical technique (interval and ratio measures are grouped together because, as noted in Chapter 3, practically all analysis methods suitable for ratio measures are also applicable to interval measures).

parametric statistics

non-parametric statistics

HINT 6.1

Try to work with probabilistically-drawn samples of sufficient size. You will have a greater choice of techniques (and fewer headaches) when it comes to analysis.

univariate analysis

bivariate analysis

Variable 2	Variable 1		
	Nominal	Ordinal	Interval/ratio
Nominal	I	II	III
Ordinal		IV	V
Interval/ratio			VI

Table 6.1 Levels of measurement and bivariate analysis

If you thought the case of bivariate analysis was bad enough, then **multivariate analysis** (i.e. taking three or more variables at a time) can only be described as positively nasty. Before you panic (and set this book on fire), however, let us assure you that we will not be dealing with multivariate statistical procedures in this masterpiece (other than providing you with a brief introduction of them in Chapter 14). Learning about multi-

multivariate analysis

variate analysis (and living to talk about it) presupposes a sound background with basic statistical concepts and familiarity with both univariate and bivariate techniques; as, at this stage, you are likely to possess neither (otherwise why did you buy/borrow/steal this book?) we shall forget about multivariate techniques for the moment.

While the level of measurement and the number of variables involved impact upon the choice of technique, the latter is also influenced by the *nature* of the measures concerned, that is whether they are independent or related. With **independent measurements**, different units (i.e. sample elements) are compared on a given characteristic. For example, contrasting the beer-drinking capacity of Liverpool football supporters with that of Bayern München supporters is an analysis example involving independent measures (i.e. the measurement of beer-drinking capacity of Liverpool supporters does not affect the measurement of the beer-drinking capacity of the Bayern München supporters and vice versa). With **related measurements**, on the other hand, the same units are compared on different characteristics. For example, contrasting the beer-drinking and whisky-drinking capacities of a sample of Liverpool supporters is an example involving related measures (since the measurements obtained on each characteristic cannot be assumed to be independent from the measurements obtained on the other as *both* measures relate to the *same* units).

Typical analysis situations involving related measures arise from longitudinal research designs and experimental studies of the 'before–after' variety (see Chapter 1). However, even with cross-sectional data, related measures may be involved, for example, when comparing the opinions of the same group on two different attitude scales. Note that in the literature the issue of measure independence is often referred to as **sample independence** (i.e. a distinction is drawn between independent and related samples rather than measures). We prefer to talk about measures rather than samples because it is not the sample elements themselves that are measured but rather their characteristics (see Chapter 3).

One practical way of checking the question of independence is to visualize the data matrix and ask the question: Are we comparing different groups of respondents (i.e. units of analysis) on a given characteristic or different characteristics of the same respondents? In the first instance, as Figure 6.2 shows, we are partitioning the data matrix by rows (i.e. splitting it into different sub-samples reflecting independent observations). In the latter instance we are partitioning the data matrix by columns (i.e. into different variables relating to the same set of observations). In the atrociously simplified example of Figure 6.2, the two partitioning approaches result in the following very different analyses: first, a comparison of supermarkets (A) and department stores (B) in terms of their *total* aftershave sales and, secondly, a comparison of *Macho Man* (A) and *Super Whimp* (B) sales irrespective of store type. In the first case, we are dealing with independent measures as we are comparing two distinct sub-samples (i.e. four supermarkets and three department stores). In the second case, we are comparing two variables across all seven stores. Different statistical techniques are appropriate for each type of comparison and you will learn all about them in Chapter 12.

*independent
measurements*

related measurements

sample independence

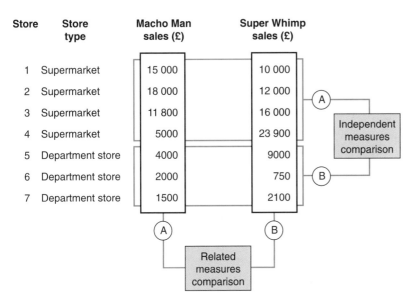

Store	Store type	Macho Man sales (£)	Super Whimp sales (£)
1	Supermarket	15 000	10 000
2	Supermarket	18 000	12 000
3	Supermarket	11 800	16 000
4	Supermarket	5000	23 900
5	Department store	4000	9000
6	Department store	2000	750
7	Department store	1500	2100

Figure 6.2   Independent v. related measures analysis.

The final set of influences affecting the choice of analysis technique has to do with the responses themselves, i.e. the actual data values obtained. A key issue here is that of item non-response, that is missing values on one or more variables entering the analysis (see Chapter 4 to refresh your memory as to why missing values may turn up and what you can do about them). A direct consequence of missing values is a reduction in the **effective sample size**, i.e. the number of observations or data points actually available for analysis; this, in turn, will affect what kind of statistical procedures can legitimately be employed (see previous discussion on the influence of sample size).

effective sample size

To illustrate the potential severity of the missing value problem, consider a situation where you have a sample of 100 roller-skating freaks (sorry, enthusiasts) and you have asked them two questions, namely (1) how many pairs of roller skates they own, and (2) how many hours per week they spend roller skating. Now, assume that, for one reason or another, you end up having 10% item non-response on each question. You may think that, given this non-response pattern, a total of 90 cases will be available for analysis. However, this is not necessarily the case, since it is possible that a *different* 10% of respondents did not answer Question 2 from the 10% that did not answer Question 1. This means that in any analysis including both questions, your effective sample size may be as low as 80 cases (depending upon the degree of overlap among non-respondents on the two questions). To make matters worse, even when you take one variable at a time, you still have the problem of **comparability of responses** (since the 90 respondents who answered Question 1 may include some or all of the 10 respondents who did not answer Question 2 and vice versa). Extend this line of thinking to a situation of many variables (i.e. a real-life data set) and you can imagine the kind of horrors that missing values can cause!

comparability of responses

A second consideration concerns the assumptions made regarding the *distribution of values in the population*. Many inferential statistical pro-

normal distribution

cedures assume that the values of the variable(s) of interest in the underlying population are 'normally' distributed, i.e. that they form a symmetrical, bell-shaped curve (this curve is known as the **normal distribution** and you will make its acquaintance in later chapters). 'Violations of the normality assumption' (an expression widely used among hard-core statisticians) can render a great deal of parametric statistical techniques inoperative. However, 'departures from normality', (another favourite expression among the statistical 'in'-crowd) have to be rather severe, since (mercifully) many statistical tests are quite *robust* (i.e. they do not immediately break-down in tears with the slightest violation in their assumptions). Nevertheless, if the assumption of normality cannot be reasonably entertained (and there are tests one can use to check for this, as we shall see in Chapter 11), then one may have to make do with non-parametric statistical tests. The latter make no assumption about the distribution of values in the population and are, therefore, also sometimes referred to as 'distribution-free' statistics. Having said all this, if the normality assumption cannot be satisfied but the sample is large enough, one may still be able to use some parametric statistical procedures (the reasons for this are quite complex to explain here, so we will not even try).

From the above, it should be obvious that a recurrent theme in the choice of analytical technique is deciding between parametric and non-parametric methods of analysis. Figure 6.3 should make this task easier for you by providing a summary of the major considerations (dear, dear, are we spoiling you or what?)

As a final point, it would be highly immoral not to mention that, in addition to the various influences displayed in Figure 6.3, different statistical procedures have some 'unique' assumptions/requirements that must also be satisfied in order for them to yield valid results (yes, we know that this makes things even more complicated but it is not our fault – blame the statisticians if you must). However, we thought that it would be better to

Figure 6.3 Choosing between parametric and non-parametric procedures.

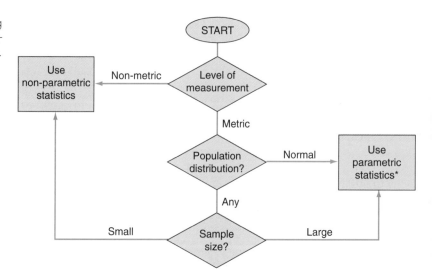

*Assuming additional assumptions are satisfied

mention such additional assumptions on an 'as needed' basis (i.e. when discussing specific procedures) rather than overload an already demanding chapter.

Summary

The purpose of this chapter was to provide an overview of the kinds of issues that must be considered *before* the analysis can be carried out. We started by stressing the importance of setting analysis objectives and followed this up by showing you how the overall research aims can be used to this effect. Next, we distinguished among different foci of analysis as reflected in description, estimation and hypothesis-testing, respectively. Last, but not least, we took a hard look at the various factors that influence the choice of analytical procedures, highlighting in particular the conditions favouring the use of parametric v. non-parametric statistics. We are now ready to actually do some analysis.

Questions and Problems

1. What are the three main reasons for setting clear analysis objectives?

2. Describe the connection between analysis objectives and research purpose. Use an example to illustrate your answer.

3. Why is it useful to distinguish between content and focus issues when setting analysis objectives?

4. Construct an example to illustrate the differences between description, estimation and hypothesis-testing.

5. In simple terms, what is the purpose of inferential statistics?

6. Which key factors, other than the objectives of analysis, determine the choice of statistical technique?

7. When would you use parametric versus non-parametric statistics?

8. What are the implications of item non-response for data analysis?

9. Draw up an example to illustrate the difference between independent and related measures.

10. Wouldn't life be simpler without Figure 6.1?

Notes

1. Stockton, JR & Clark, CT (1975) *Introduction to Business and Economic Statistics*, 5th edn, p. 14. Cincinnati, OH: South-Western Publishing Co.

Further Reading

Sorry guys, there is nothing useful we know of to refer you to for your bed-time reading. For some reason, and despite their importance, the issues we covered in this chapter have not received explicit and integrated treatment in mainstream data analysis textbooks.

PART III:

CARRYING OUT THE ANALYSIS

Why not take it easy initially and *describe* your data?

Purposes of Data Description

Imagine that you have just completed a nationwide survey of Eskimo attitudes towards year-round swimming. Your research instrument, a 24-page mail questionnaire, requested information on some 120 different variables ranging from sociodemographic characteristics (e.g. age, education and marital status) to behavioural aspects (e.g. number of times that respondent went swimming in the sea during December). Out of 2000 questionnaires initially sent out, you managed to get back 350 fully completed (as well as 868 letters calling you a lunatic, 231 threatening phone calls from outraged respondents, 12 personal visits from respondents who 'did not see the funny side', and three letter bombs). Following your recovery from hospital, you have coded your data and are now facing something like 42 000 entries in your data matrix (350 cases times 120 variables). Now, no matter how brilliant you are or how long you spend staring at the data matrix, you will be unable to make any sense out of the endless list of seemingly random numbers. This is not surprising, since

> Simple inspection of a collection of numbers will ordinarily communicate very little to the understanding of the investigator. Some form of classification and description of the numbers is required to assist interpretation and to enable the information which the numbers contain to emerge.[1]

Data description is a typical first step in any data analysis project. In addition to being an important, self-standing activity when a descriptive focus characterizes the analysis objectives (see Chapter 6), descriptive analysis provides a very useful initial examination of the data even when the ultimate concern of the investigator is inferential in nature (i.e. involving estimation and/or hypothesis-testing). Specifically, the purpose of descriptive analysis is to:

(a)   Provide preliminary insights as to the nature of the responses obtained, as reflected in the distribution of values for each variable of interest.

(b)   Help detect errors in the coding process (see also Chapter 4).

(c)   Provide a means for presenting the data in a digestible manner, through the use of tables and graphs.

frequency distribution

absolute frequencies
relative frequencies

cumulative frequency
distributions

(d)   Provide summary measures of 'typical' or 'average' responses as well as the extent of variation in responses for a given variable (to be discussed in Chapter 8).

(e)   Provide an early opportunity for checking whether the distributional assumptions of subsequent statistical tests are likely to be satisfied (see also Chapters 6 and 8).

In short, data description is like a first date: it enables you to get to know your data before you try something more adventurous.

Frequency Distributions

The starting point in descriptive analysis is the construction of a **frequency distribution** for each variable of interest. This simply shows in absolute or relative (i.e. percentage) terms how often the different values of the variable are actually encountered in one's sample. In other words, a frequency distribution indicates how 'popular' the different values of the variable are among the units of analysis. A few examples of frequency distributions using your data on Eskimos' swimming habits are shown in Table 7.1. Both **absolute frequencies** (i.e. simple counts) and **relative frequencies** (i.e. percentages) are given and, in an attempt to spoil you even more, the variables summarized cover different levels of measurement (and if you can't tell *at a glance*, which variable gives you what level of measurement, you'd better hurry back to Chapter 3 and brush up!).

There are a few things worth noting about frequencies. First, a frequency (whether absolute or relative) can *never* be negative, as a value cannot be encountered less than zero times. Second, the sum of absolute frequencies *must* equal the total number of observations (i.e. the sample size), while the sum of relative frequencies *must* equal 100% (if expressed in percentages) or 1 (if expressed in proportions). Third, as is obvious from Table 7.1, frequency tables can be set up for any variable, irrespective of the level of measurement.

The above features can be used to your advantage because they can alert you to possible errors in your data (yes, finding errors is depressing, but not half as depressing as detecting errors after you have completed your analysis). Thus, if you find that the sum of absolute frequencies for a given variable is greater than your sample size or that you obtain frequencies for 'out-of-range' values, something has gone wrong somewhere. For example, if you find that 138 of your respondents were 86 years old and your sample consisted of 75 primary school children, chances are that some entries have been coded incorrectly (probably because you did not follow our brilliant advice on editing and coding in Chapter 4). Your frequency table will be able to tell you *where* these errors are likely to have happened so that you can go back to your original data and sort things out before carrying on with the analysis.

With ordinal-, interval- and ratio-level variables it is also possible to construct **cumulative frequency distributions**. These show in absolute or relative terms how many observations take values that are 'greater than' or 'less than' a specified value. Going back to Table 7.1 you may

Variable	Values	Absolute frequency	Relative frequency (%)
Marital status	1 (Single)	183	52.3
	2 (Married)	94	26.9
	3 (Widowed)	22	6.3
	4 (Divorced)	51	14.5
	Total:	350	100.0
Education	1 (Postgraduate degree)	68	19.4
	2 (Undergraduate degree)	121	34.6
	3 (High school diploma)	87	24.9
	4 (Primary school certificate)	50	14.3
	5 (Kindergarten only)	24	6.8
	Total:	350	100.0
'I love to go swimming in December when I get a chance'	1 (Strongly disagree)	76	21.7
	2 (Disagree)	49	14.0
	3 (Neither agree nor disagree)	12	3.4
	4 (Agree)	73	20.9
	5 (Strongly agree)	140	40.0
	Total:	350	100.0
Number of children	0	110	31.4
	1	74	21.1
	2	38	10.9
	3	15	4.3
	4	20	5.7
	5	56	16.0
	6	0	0.0
	7	13	3.7
	8	9	2.6
	9	13	3.7
	10	2	0.6
	Total:	350	100.0

Table 7.1 Examples of frequency distributions

want, for example, to find out how many respondents had three children at most or how many had more than five children. To answer these and similar kinds of questions, you can use the absolute and/or relative frequencies to compute the corresponding cumulative frequencies. This simply involves adding the frequencies associated with a particular value to the sum of the frequencies corresponding to all preceding values. Table 7.2 shows the derived cumulative frequency distributions for 'number of children' based on the information given in Table 7.1.

Table 7.2   An example of a cumulative frequency distribution[a]

Number of children	Cumulative frequency	Cumulative relative frequency (%)
0	110	31.4
1	184	52.5
2	222	63.4
3	237	67.7
4	257	73.4
5	313	89.4
6	313	89.4
7	326	93.1
8	335	95.7
9	348	99.4
10	350	100.0

Based on frequencies shown in Table 7.1.

Again, cumulative frequencies (whether absolute or relative) must be non-negative. Moreover, the last absolute cumulative frequency must be equal to the sample size, while the last relative cumulative frequency must be equal to 100% (see Table 7.2); if it's not, you'd better check your data!

Cumulative frequency distributions such as the ones shown in Table 7.2 are of the 'up to' or 'less than' variety, that is they tell us the number/percentage of observations *not exceeding* a certain value (e.g. that 237 (67.7%) respondents had three children or fewer). It is very easy, however, to calculate the number/percentage of observations that *exceed* a given value (e.g. in order to find out how many respondents had more than three children). This is quickly done by subtracting the absolute cumulative frequencies from the sample size or by subtracting the relative cumulative frequencies from 100%. In our example, 350–237=113 respondents had more than three children; this represents 100%–67.7%=32.3% of all respondents.

percentiles

A particularly useful application of cumulative frequencies is the calculation of what are known as **percentiles**. These divide a set of observations into 100 equal portions; thus, there are 99 percentiles ($P_1, P_2, P_3, \ldots, P_{99}$). The pth percentile P_p is the point on the measurement scale such that p percent of the observations have values less than P_p and $(100-p)$ of the observations have values greater than P_p. For example, the 25th per-

first quartile

centile P_{25} (also known among statistical boffins as the bottom or **first quartile**, Q_1) is the value below which 25% of the observations fall; the

second quartile

top quartile

50th percentile P_{50} (the **second quartile**, Q_2) is the value below which 50% of the observations fall; and the 75th percentile P_{75} (the **top quartile**, Q_3) is the value below which 75% of the observations fall. By looking at percentiles we can get a feel of how our sample is spread along the variable of interest. Thus knowing that $P_5=10.7$ and $P_{15}=20.5$ tells us (a) that 5% of our sample have values less than 10.7, (b) that 15% of the sample have values less than 20.5, and (c) that 10% of the observations lie between 10.7 and 20.5. Similarly, knowing that $P_{90}=26.4$ indicates that only 10% of the observations have values greater than 26.4, while 90% lie below.

As an example, say that we want to establish the 80th percentile (P_{80}) for the 'number of children' variable in Table 7.2. In other words, we want to find out the value below which $(80/100) \times 350 = 280$ of the observations fall. Looking at the absolute cumulative frequencies we can see that 257 respondents had 4 children or less; this leaves us short of 23 respondents $(280 - 257)$. To get these we must 'travel' a distance of $(23/56) = 0.41$ into the next category to make up the shortfall (the next category containing $313 - 257 = 56$ individuals with 5 children – see also Table 7.1). So the 80th percentile is somewhere between 4 and 5 children; in fact, strictly speaking $P_{80} = 4 + 0.41 = 4.41$, which indicates that 80% of the respondents had less than 4.41 children.

Obviously, in this particular example, the exact value for the 80th percentile is not very illuminating, as '4.41 children' does not make much sense. This is because, being a **discrete variable**, 'number of children' can only be measured in whole units; in this context, we might just as well have read directly from Table 7.2 the percentages of respondents with a maximum of 4 children and/or 5 children (which come to 73.4% and 89.4%, respectively). However, had we been dealing with a **continuous variable**, that is one that can assume (at least theoretically) any value within an interval (e.g. age, weight, height, income) then all percentiles would be meaningful (as all values, fractional or not would be legitimate). Note that the procedure for finding a desired percentile would be identical to the one followed above (try it out on a continuous variable of your choice, if you don't believe us).

discrete variable

continuous variable

Grouped Frequency Distributions

Sometimes, it is not practical to construct frequency distributions based on the original values of a given variable. With a continuous variable or with a discrete variable that takes on many individual values, deriving a frequency distribution along the lines described previously is unlikely to be very informative. Consider for example, the variable 'age' in the context of the Eskimo swimming survey. Being a continuous variable, in the extreme case, the age of each of the 350 respondents is likely to be different from that of everyone else's (unless two or more respondents happened to be born exactly at the same time – i.e. down to a nanosecond!). A frequency distribution constructed under this (admittedly, extreme) scenario would convey nil additional information over the original data as there would be 350 individual frequencies – with each frequency being equal to 1! Even if we had measured the 'true' age only approximately (e.g. by asking respondents to indicate their 'age at nearest birthday'), there is still likely to be a very large number of values to deal with. For example, even if the survey was confined to adults (i.e. 18-year-olds and over) and the oldest respondent was only 87 years young, not fewer than 69 different yet perfectly legitimate responses (i.e. ages) would be contained in this range (including the youngest and oldest respondents). Again, a frequency distribution with 69 categories is not going to be much help for interpretation purposes.

As you may have guessed, the solution to problems posited by too many values is to group the latter into a smaller number of **classes**. Each class or **class interval** is defined by setting two **class limits** (an upper and a

class

class interval, class limits

Table 7.3 An example of a grouped frequency distribution

Age of respondent	Absolute frequency	Relative frequency	Cumulative frequency (%)
18–24	67	19.1	19.1
25–34	89	25.4	44.5
35–44	106	30.3	74.8
45–54	33	9.4	84.2
55–64	28	8.0	92.2
65–74	23	6.6	98.8
75 and older	4	1.2	100.00

lower) within which a subset of the original values is subsumed. Having grouped the original values into appropriate class intervals (we will see how this is done shortly), one can proceed to derive a **grouped frequency distribution**, which shows how often the different classes of the variable concerned are encountered in the sample. Table 7.3 shows the grouped frequency distribution for our age example.

grouped frequency distribution

Note that class limits are set in a *non-overlapping fashion* (i.e. not 18–25, 25–35, etc., but 18–24, 25–34, etc.); had overlapping limits been set, then a respondent who was exactly 25 years old could have been allocated to two different classes. This would violate a key classification principle, notably that of having **mutually exclusive** categories. Another key classification principle is that of **collectively exhaustive** categories, which says that you cannot have any left-overs at the end – every observation has to fit somewhere.

mutually exclusive
collectively exhaustive

WARNING 7.2

When setting class intervals, make sure that they are not overlapping and that all cases can be assigned to a class.

Class limits reflect the nature and extent of rounding-off of the figures. The ages in the present example are all rounded off to the *nearest* year (since it was age at the nearest birthday that was measured). Thus a respondent between 24 and 24.5 years old would be placed in the first class interval, while a respondent between 24.5 and 25 would be counted in the second class interval. This implies that the 'true' boundary between the first and second class intervals is 24.5. Similar 'true' or 'exact' boundaries can be determined between the second and third intervals (34.5), the third and fourth (44.5), and so on. These are known as **true class limits** (or simply 'class boundaries') to distinguish them from the (non-overlapping) **stated class limits** typically used in displays of grouped frequency distributions. In general, when data are recorded to the nearest unit (as is mostly done), the lower true limit of any given class lies one-half unit below the lower stated limit and the upper true limit lies one-half unit above the upper stated limit. Thus the true class limits of the first three intervals in Table 7.3 are 17.5–24.5, 24.5–34.5, and 34.5–44.5, respectively.

true class limits
stated class limits

By using the true class limits we can establish the **class width** (which shows the number of measurement units included in the class, i.e. the size of the class interval) and the **class midpoint** (which can be thought of as the 'representative' value of the class). Specifically, the class width is defined as

class width

class midpoint

$$w = U - L$$

(where U and L are the *true* upper and lower limits, respectively) while the class midpoint (M) is defined as

$$M = L + 0.5w = U - 0.5w = \frac{U + L}{2}$$

In the example shown in Table 7.3, the width of the first class interval comes to $24.5 - 17.5 = 7$ and the class midpoint is, therefore, $17.5 + (0.5 \times 7) = 21$. Similarly, the width of the second class is $34.5 - 24.5 = 10$ and the class mid-point $24.5 + (0.5 \times 10) = 29.5$. Just as a piece of useless information, if the width of the interval is odd (i.e. the interval contains an odd number of units), the class midpoint will be a whole number; if the interval width is even, the class midpoint will be a fraction.

Clearly, the distinction between stated and true class limits only makes sense for continuous variables. With *discrete* variables, the stated and true class limits are identical as only whole units are permissible as values. For example, if 'number of children' is grouped into three classes 0–3, 4–7 and 8–10, there are no legitimate values between, say, the upper stated limit of the first class and the lower stated limit of the second class: someone can have *either* 3 *or* 4 children with no other possibility in between (see also previous discussion on percentiles).

There are a few words of wisdom to bear in mind when setting up grouped frequency distributions. First, it goes without saying that by grouping values together there is, inevitably, loss of information. Thus, from Table 7.3, it is not possible to identify the individual ages of the 89 respondents who are between 25 and 34 years old; some of them may be 25, others 27 and so on. We could take the class midpoint as a 'representative' value (and this is what is mostly done in practice), but by doing this we implicitly assume that responses are evenly distributed within the class interval (or that all responses are the same and equal to the midpoint value). If either is roughly the case, then there is no harm in using the class midpoint to represent a given class. If, on the other hand, responses are heavily concentrated in the upper or lower parts of the interval, then using the class midpoint may result in a distorted picture of the data. Of course, we could go back to the original (i.e. ungrouped) data and see how responses are actually distributed within the interval(s) concerned. However, this assumes that we *can* go back to the original data, that is that access is not a problem (e.g. that we have not torn/burned our question-naires and/or coding sheets while throwing a tantrum).

Second, for a given set of data, the loss of information becomes larger as the class width increases and the number of classes decreases. Had we, for example, grouped all ages between 18 and 34 in a single interval, there would be 156 respondents whose ages might vary by as much as 17 years from each other; at the same time, the number of classes would have been reduced from seven to six. So the dilemma one faces is that 'using too few intervals results in an excessive loss of information. Using too many defeats the purpose of summarization'.[2]

Fortunately for you, there are some general guidelines to help you con-struct grouped frequency distributions that provide an effective summary of your data without incurring an unacceptable loss of information; these are gracefully outlined in Table 7.4.

Table 7.4 Guidelines for grouping data

- Categories should be mutually exclusive.
- Responses within categories should be similar.
- Substantive differences in responses should exist between categories.
- Categories should be exhaustive.
- The number of categories should be neither too large nor too small (say, between 6 and 15 categories).
- Preferably, categories should be of equal width.
- Class intevals of 5, 10 or some multiple of 10 units tend to be easier to comprehend.
- Open-ended categories are to be avoided if possible.

There are also a couple of weird formulae that may help you decide on the number of class intervals and their widths and, thus, demonstrate your unparalleled statistical grouping skills to friends and family alike. These formulae come in particularly handy when you are dealing with a continuous variable and a large sample (e.g. if you recorded the exact weight of 1200 stray cats during your last holiday in Greece).

Deciding on the 'optimal' number of categories (c) can be accomplished by following Sturges' famous rule:

$$c = 1 + 3.322(\log_{10} n), \qquad \text{where } n = \text{sample size}$$

Although this is not a hard and fast rule to be followed uncritically in all situations (see additional considerations in Table 7.4), it can be quite helpful. For example, application of Sturges' rule suggests that we may have used too few class intervals in grouping the respondents' ages in Table 7.3, resulting perhaps in too crude a classification (about 9 intervals would have been 'optimal' according to this rule).

Having decided on the number of intervals, it is a simple matter to get an approximation of the class width (w) you should be aiming for. This is given by

$$w = \frac{\text{max} - \text{min}}{c}$$

(where max and min are the highest and lowest values, respectively). In our example, the youngest respondent was 18 years old and the oldest 87, so $w = (87 - 18)/9 = 7.66 \approx 8$ (based on 9 intervals). Again, the formula for w should not be followed blindly as the resulting class width may be inconvenient to work with or not customarily used for the variable under consideration. For example, ending up with a class width of 18 to record income levels (e.g. 1000–18 000, 19 000–36 000, 37 000–54 000, etc.) is not nearly as appetizing as, say, a class width of 20 (i.e. 1000–20 000, 21 000–40 000, 41 000–60 000, etc.). As the more observant amongst you (or those still awake) may have noticed, the above class width has been calculated using the true class limits, as income is a continuous variable.

As Table 7.4 recommends, ideally, all class intervals should be of the same width. This makes life much easier both for comparing the frequencies of the various classes and for graphical display purposes (to be discussed shortly). However, if the nature of the data is such that there is a large number of values relatively close to each other, coupled with a few values very far apart, then one may be forced to use **unequal class intervals**

unequal class intervals

(i.e. narrower intervals for the former set of values and wider intervals for the latter set). For example, in Table 7.3, the first (18–24) and second (25–34) intervals are of unequal width, with the former spanning a narrower range of ages than the latter.

Sometimes **open-ended intervals** are used at either or both ends of a grouped frequency distribution (however, as with unequal class intervals, they should be avoided if at all possible). These are typically employed when there are a few extremely small or extremely large values in the data in order to avoid having one or more class intervals with zero frequencies. A rather nasty problem with open-ended intervals is that their widths are not readily apparent. For example, looking at the (grouped) age distribution in Table 7.3, there is no way of knowing how close or far apart the four 'veteran' year-round swimmers (75+) are in terms of their age. The inability to determine class width, leads to a second headache associated with open-ended intervals, namely that of identifying a 'representative' value (as a 'proper' class midpoint does not exist). One must either go back to the original (ungrouped) data (assuming this is feasible), or take a chance by guessing a midpoint (e.g. by making the assumption that the class width is the same as for the other intervals; however, if unequal class widths have been used for the latter, things become very messy indeed as there is no obvious choice).

By now you must be getting the rather uncomfortable feeling that, while grouped frequency distributions are helpful for *summarizing* a mass of data, there is a price to pay when it comes to making *calculations* based on such data. Questions of distributions of observations within class intervals, unequal class widths and the problems of open-ended intervals can all introduce error in subsequent analysis, such as the computation of summary measures (e.g. averages) for the variables concerned. While it is possible to calculate such summary measures based on grouped data (and there are plenty of horrible-looking formulae which you can find, if you must, in the Further Reading section), our position on this issue is simple: by all means construct grouped frequency distributions to present your data in a digestible fashion; even better, make good use of graphs and diagrams (advice on which will follow shortly) to improve on the effectiveness of your presentation. However, always use the original (ungrouped) data when computing measures of average and/or dispersion so as to avoid potential inaccuracies introduced by the grouping process (we will discuss a variety of summary measures in Chapter 8).

open-ended intervals

HINT 7.2

As the grouping of data into categories results, inevitably, in some loss of information, calculation of summary measures based on *ungrouped* data is preferable.

Graphical Representation of Frequency Distributions

It is often said that 'a picture speaks a thousand words' and this is certainly the case with frequency distributions. Rather than relying solely on frequency tables to describe your data, you may decide to go for a **graphical representation**, if only to impress the reader with your amazing ability to draw pretty pictures or show off the extortionately expensive graphics package you recently bought for your computer. On a more serious note, using graphical representations is almost always a good idea, not least because 'the average casual reader is likely to give scant attention to the

graphical representation

ordinary printed matter in a research report . . . However, his eye is likely to be arrested by any picture or chart that may happen to be included, and this may lead him to read the entire discussion'[3] (see also Chapter 15 on presenting the analysis).

While the tabular form of representing frequency distributions can be used with any variable irrespective of its level of measurement (see the various examples in Table 7.1), things get slightly more complicated if you want to be 'artistic'. This is because some types of graphs are better suited to one level of measurement than another.

bar chart, pie chart

If you are dealing with a nominal or ordinal variable, then you cannot go wrong with a **bar chart** or a **pie chart**. In the former, values are represented by vertical or horizontal bars the height of which is proportional to each value's absolute or relative frequency. In a pie chart, a circle is divided into slices representing the various values with the size (i.e. area) of each slice being proportional to the value's *relative* frequency (pie charts are not really used to display absolute frequencies). Figure 7.1 shows the information on 'marital status' and 'education' from Table 7.1 in bar chart and pie chart form, respectively.

Figure 7.1   Examples of bar and pie charts.

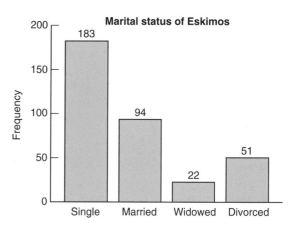

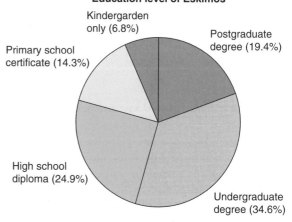

With interval and ratio data, the choice of graph depends partly on whether the variable concerned is discrete or continuous and partly on whether the data is grouped or not. With a discrete variable that takes only few values, you can use the same type of display as for nominal and ordinal variables (i.e. bar charts and/or pie charts); the 5-point Likert scale used to register attitudes towards swimming in Table 7.1 is a prime example here. With discrete variables which take a lot of different values and with continuous variables which are in the form of grouped frequency distributions, one has the choice between a **histogram** and a **frequency polygon**. In a histogram, absolute or relative frequencies are represented by areas in the form of bars. The total area under the histogram is 1.0 (or 100%) so the proportion of the area over a range of values shows the proportion of observations falling into that range. The bars are erected at either the true limits of each class interval (the 'purist' approach) or the stated limits (the 'practical' approach); in both cases, adjacent bars 'touch' each other, unlike in bar charts where the bars are set apart. Also unlike bar charts, the height of each bar is *not* proportional to a value's frequency; rather it is the *area* of the histogram bar that reflects the latter. Consequently, unless all class intervals are of equal width (in which case heights and areas are in constant proportions), it is not meaningful to have a frequency scale on the vertical axis (incidentally, statistics texts often display histograms with a vertical (frequency) scale, but few make explicit the assumption concerning equality of intervals – so beware).

Now, if you connect the bars of the histogram at the midpoints of the class intervals, you will magically end up with the second type of graph, namely the frequency polygon. Figure 7.2 shows the histogram and frequency polygon for the (grouped) age data contained in Table 7.3; the figures at the top of each histogram bar indicate the percentage of cases falling within each age class (see Table 7.3).

**histogram
frequency polygon**

WARNING 7.3

The heights of bars in a histogram are not proportional to frequencies *unless* all class intervals are of exactly the same width.

Age of Eskimos

19.1%   30.3%
    25.4%

9.4%

8.0%
    6.6%

1.2%

18–24 25–34 35–44 45–54 55–64 65–74 75–84

Figure 7.2 Examples of a histogram and a frequency polygon.

Note that the open-ended interval at the top end of the age distribution in Figure 7.2 has been replaced by a closed interval (75–84) to enable the construction of the complete histogram (otherwise the last class could not have been represented by a histogram bar). Note also that the frequency polygon is closed, by convention, by connecting the midpoint of the first (last) interval with a point on the horizontal axis one-half a class interval below (above) the lower (upper) true limit of the first (last) class (if this sounds a bit complicated, have another look at Figure 7.2 and things should become clear).

Histograms and frequency polygons are based on slightly different assumptions regarding the distribution of observations within each class interval. In a histogram, frequencies are represented as being equally distributed over the range of a given interval, and changes from interval to interval are shown in a stepwise fashion. In contrast, with a frequency polygon we assume that all cases within an interval are concentrated at the midpoint of the interval and, thus, changes from interval to interval are depicted as continuous. In fact, if we imagine a situation whereby (equal) class intervals become smaller and smaller and the number of observations (i.e. sample size) becomes larger and larger, we would eventually end up with an infinitely small class interval and an infinitely large number of cases (see Figure 7.3(a)), i.e. a truly **continuous frequency distribution**. The frequency polygon of the latter would then majestically show up as a smooth curve (see Figure 7.3(b)). This type of representation is often used to graphically depict a frequency distribution in general terms and we will be making use of it in subsequent chapters.

continuous frequency distribution

Figure 7.3 Graphical representation of a continuous frequency distribution. (a) Histogram; (b) frequency polygon.

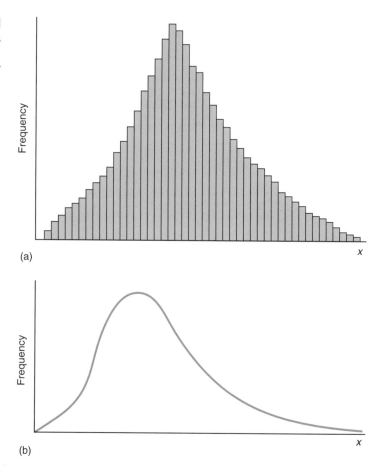

One advantage of the frequency polygon over the histogram is that when one wants to plot two (or more) distributions that overlap on the same base line (i.e. horizontal axis), the histogram can give a rather messy and confused picture. In contrast, the frequency polygon, being simpler, usually enables a better comparison. An illustration indicating the final examination scores of two groups of meteorology students is shown in Figure 7.4; note that since the data are recorded in equal intervals, a frequency scale is given on the vertical axis of the histograms and frequency polygons to aid interpretation.

By now, you must be getting extremely anxious to learn what kind of graph one may use to depict *cumulative* frequency distributions; after all, the various types of graphs we have discussed so far only apply to absolute or relative frequencies. One way of doing this is by means of the **cumulative frequency polygon** (what a mouthful), which differs from the 'normal' frequency polygon in two respects. First, instead of plotting points corresponding to absolute frequencies, we plot points corresponding to

cumulative frequency polygon

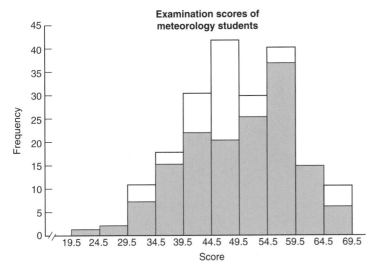

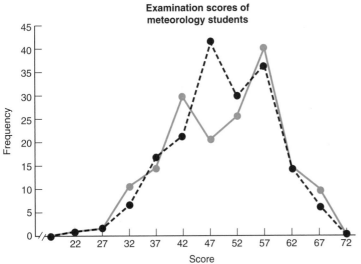

Figure 7.4 Graphical comparison of frequency distributions.

HINT 7.3

If you want to graphically compare frequency distributions on the same scale, a plot of their frequency polygons will give a tidier picture than a plot of their histograms.

cumulative frequencies (you would have never guessed, would you?). Second, instead of plotting points above the midpoint of each class interval, we plot points above the upper true limit of each class (so that the graph shows the number of observations falling above/below particular values).

Figure 7.5(a) shows the cumulative frequency polygon for the data in Table 7.2. Note that since 'number of children' is a discrete variable, the stated and true class limits are identical (remember?) and, given that the data has not been grouped, each 'class interval' has a width of one unit. Note also that the general trend of the cumulative frequency polygon is progressively rising; there are no inversions or setbacks. This is because all (non-cumulative) frequencies are positive except the odd zero frequency (e.g. as in the case of 6 children).

ogive   A more useful type of graph (yes, we know you have had enough of graphs – this *really* is the last one) is the **ogive**, which plots the cumulative relative frequencies against the variable's values (see Figure 7.5(b)). You can use the ogive to (a) name your dog (hey, Ogive, here boy!), (b) read off directly the percentage of observations less than any specified value, and (c) determine the value below which a specified percentage of observations lies. Thus in Figure 7.5(b), if you were interested in finding out the per-

Figure 7.5 Cumulative frequency polygon (a) and ogive (b).

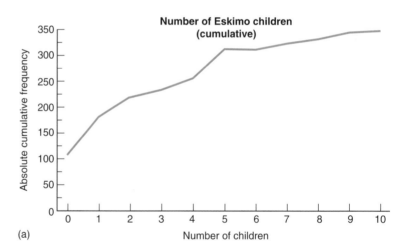

(a)

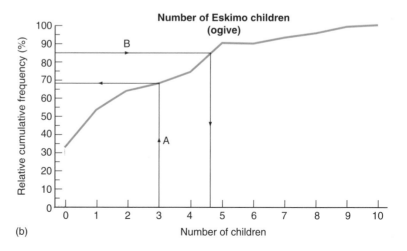

(b)

centage of respondents that had 3 children at most, you would follow arrow A; this would tell you that almost 70% of all respondents had between 0 and 3 children. If, on the other hand, you wanted to know how many children 85% of the respondents had, following arrow B would give you the answer 'fewer than 5 children'. Thus the ogive can be used as an alternative (and quick) way for looking at percentiles.

Most statistical analysis packages (see Chapter 5) have facilities for producing the various kinds of frequency distributions and graphical displays we have discussed in this chapter (and usually many more); this means that the only thing you have to do is input your data and the computer does the rest. Table 7.5 shows an example of the kind of output you can expect from a decent statistical package. Here we are using our favourite analysis package, *SPSS/PC+*, on data relating to attitudes towards winter swimming by a sub-sample of 100 Eskimos (see 5-point Likert scale in Table 7.1).

Table 7.5 Example of SPSS/PC+ frequencies output

WINTERSWIM Attitude Toward Winter Swimming

Value label	Value	Frequency	Percent	Valid percent	Cumulative percent
Strongly disagree	1	7	7.0	7.1	7.1
Disagree	2	6	6.0	6.1	13.3
Neither agree nor disagree	3	19	19.0	19.4	32.7
Agree	4	36	36.0	36.7	69.4
Strongly agree	5	30	30.0	30.6	100.0
Missing	9	2	2.0	Missing	
	Total	100	100.0	100.0	

```
1 \\\\\\\ 7
2 \\\\\\ 6
3 \\\\\\\\\\\\\\\\\\\ 19
4 \\\\\\\\\\\\\\\\\\\\\\\\\\\\\\\\\\\\\ 36
5 \\\\\\\\\\\\\\\\\\\\\\\\\\\\\\ 30
  +----+----+----+----+
  0    8   16   24   32
```

Percentile	Value	Percentile	Value	Percentile	Value
5.00	1.000	10.00	2.000	25.00	3.000
50.00	4.000	75.00	5.000	95.00	5.000

Valid cases 98   Missing cases 2

Finally, when using the graphics options of statistical software or employing a specialized graphics package, it is worth bearing in mind some general guidelines regarding graph construction (as sole reliance on 'default' settings may result in boring or even weird-looking graphs!). A set of such guidelines is generously provided in Table 7.6.

Table 7.6   Guidelines for graph construction

- Every graph should have a title (either above or below the diagram).
- The general arrangement of a graph should proceed from left to right.
- Where possible, quantities should be represented by linear magnitudes as areas or volumes are more likely to be misinterpreted.
- The horizontal scale for a graph should read from left to right and the vertical scale from bottom to top. Low numbers on the horizontal scale should be on the left and low numbers on the vertical scale should be towards the bottom.
- Both the horizontal and the vertical axis should be clearly labelled. In graphing frequency distributions, it is customary to let the horizontal axis represent values and the vertical axis frequencies.
- Preferably, the zero point should be included in the graph (e.g. for a frequency scale on the vertical axis); if this is not practicable, a break in the axis should be used.
- While the distance on each axis chosen to serve as a unit is arbitrary, it can affect the appearance of a graph; a rough 3:5 height to length ratio seems to work well in most cases.
- The coordinate axes for a graph should be sharply distinguished from any other lines included.
- When representing percentages, it is helpful to emphasize in some distinctive way the 100% line (or any other basis of comparison).
- The graph should not be overly cluttered with too much information, but the key information should be made as clear as possible.

Summary

In this chapter, we took the first steps in data analysis by attempting to describe a set of data. We started by looking at the concept of a frequency distribution and the various forms it can take. We then focused on the process of grouping data and the potential problems that may be encountered along the way. Finally, we let our artistic talents run wild by examining different types of graphical displays. We are now ready to start some serious work, namely to derive summary measures that can be used to succinctly represent the properties of a set of data.

Questions and Problems

1. What is the role of data description?

2. What is the difference between an absolute and a relative frequency distribution?

3. Why is it not possible to calculate a cumulative frequency distribution for nominal-level variables?

4. What are percentiles?

5. What is the difference between a discrete and a continuous variable?

6. What is the difference between 'true' and 'stated' class limits?

7. Which factors should be considered when constructing grouped frequency distributions?

8. Why should unequal class intervals and open-ended intervals be avoided, if possible?

9. When would you use a frequency polygon over a histogram?

10. Do you think Ogive is a good name for a dog?

Notes

1. Ferguson, GA (1981) *Statistical Analysis in Psychology and Education*, 5th edn, p. 17. London: McGraw-Hill.
2. Daniel, WW & Terrell, JC (1992) *Business Statistics*, 6th edn, p. 15. Boston: Houghton Mifflin.
3. Glass, GV & Stanley, JC (1970) *Statistical Methods in Psychology and Education*, p. 38. Englewood Cliffs NJ: Prentice-Hall.

Further Reading

Hartwig, F & Dearing, BE (1980) *Exploratory Data Analysis*. London: Sage Publications. A concise guide to some alternative perspectives for exploring the structure of data; it complements the material covered in this as well as the next chapter.

Selby, PH (1979) *Using Graphs and Tables*. New York: Wiley. Its title says all.

Vogt, WP (1993) *Dictionary of Statistics and Methodology: A Non-Technical Guide for the Social Sciences*. London: Sage Publications. It's about time we sent you to one of those, as the amount of terminology you must digest is getting more and more; this book should help you maintain your sanity!

Waalgren, A, Wallgren, RP, Jorner, U & Haaland, J-A (1996) *Graphing Statistics & Data: Creating Better Charts*. London: Sage Publications. All you could ever hope to learn about charting; with lots of useful illustrations.

Chapter 8

Can you use few numbers in place of many to *summarize* your data?

Characterizing Frequency Distributions

In the previous chapter you learned more than you ever wanted to know about frequency distributions. Here we will build upon your knowledge by showing you how frequency distributions can be conveniently described and compared to one another with reference to some key properties they possess. These properties are important because they allow us to calculate some **summary measures** capturing the essential characteristics of different distributions. By using summary measures, we can condense the information contained in the individual values, thus making interpretation of the data much more manageable. Moreover, instead of graphically displaying distributions (as was done in Figure 7.4 in Chapter 7), we can compare their respective summary measures. Finally, as we will see in Chapter 9, we can use summary measures derived from our sample data to make inferences about the population from which the data has come.

summary measures

Consider the three distributions in Figure 8.1, representing the IQ scores of economics students enrolled at three universities, A, B and C. Although they are identical in shape, they are markedly different in terms of the central values about which the observations in each distribution appear to concentrate. In other words, the **central location** (or 'central tendency') of each distribution is different. Measures of central location are called **averages** and reflect 'middle' points in the sense that they are near the centre of the distribution. In our example distribution A has a lower average than distribution B, and B has a lower average than C. The familiar **arithmetic average** (i.e. the value obtained by adding together a set of values and dividing by their number) is one such measure; however, there are several other types of averages as we will see shortly.

central location

averages

arithmetic average

Now consider Figure 8.2, showing the distributions of the IQ scores of biochemistry students in the same three universities. Although they all have the same average, in university A the IQ scores are more closely concentrated about the average than in university B; the same applies to the IQ scores in B in relation to those at university C. Thus distributions A, B and C differ in terms of their **variability** (or 'dispersion'), that is the degree of clustering about a central value. In general, if all the observations are close to the central value (i.e. the average), their variability will be less than if they tend to depart markedly from the central value.

variability

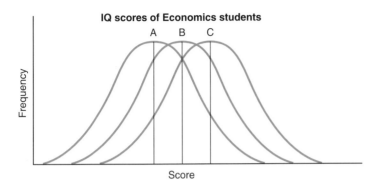

Figure 8.1   Three frequency distributions identical in shape but with different averages.

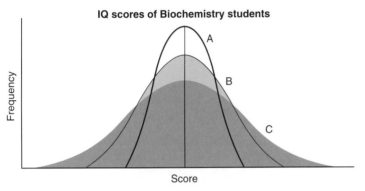

Figure 8.2   Three frequency distributions with the same average but differing in variability.

A third way in which frequency distributions can differ is in terms of their **skewness**, which reflects the 'symmetry' (or lack of it) in the distribution. Figure 8.3 illustrates this mysterious property by comparing the distributions of IQ scores of engineering students in universities A, B and C. Note that distribution B is **symmetrical**, for if we were to 'fold' it over about the average we would find that it had the same shape on both sides. In contrast, distributions A and C are **asymmetrical**, the shape to the left of the average being different from the shape to the right. More specifically, distribution A is **positively skewed** as the larger frequencies tend to be concentrated towards the low end of the variable and the smaller frequencies towards the high end (i.e. the 'tail' of the distribution extends to the right). The opposite holds true for distribution C, which is **negatively skewed** (the larger frequencies are towards the high end of the variable and the smaller frequencies towards the low end, i.e. the 'tail' is to the left).

skewness

symmetrical distribution

asymmetrical distribution

positively skewed distribution

negatively skewed distrubution

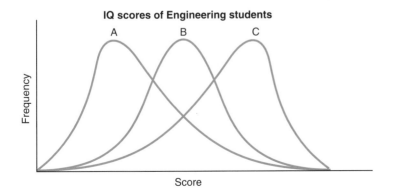

Figure 8.3   Three frequency distributions differing in skewness.

Finally, have a look at Figure 8.4, which shows the distribution of IQ scores of physics students in the three universities. These distributions differ in terms of their **kurtosis**, which describes the 'flatness' or 'peakedness' of one distribution in relation to another. If one distribution is more peaked than another, it is said to be more **leptokurtic** (compare A to B) and if it is less peaked it is said to be more **platykurtic** (compare B to C). It is conventional to speak of a distribution as leptokurtic (platykurtic) if it is more (less) peaked than a particular type of distribution known as the 'normal distribution' (to be discussed later in this chapter). The latter is spoken of as being **mesokurtic**, which means that it falls somewhere between leptokurtic and platykurtic shapes (and if this is all Greek to you, don't despair as even the authors get confused here!).

kurtosis

leptokurtic distribution
platykurtic distribution

mesokurtic distribution

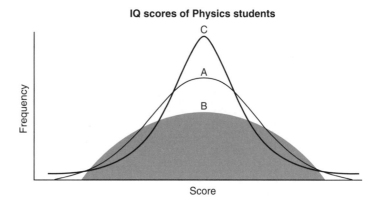

Figure 8.4 Three frequency distributions differing in kurtosis.

Taken collectively, central location, variability, skewness and kurtosis can provide a very detailed and yet efficient description of the nature of a frequency distribution. Statisticians have spent their lives developing a whole range of measures that can be used to represent these properties (well, somebody has to do it!). Here we will concentrate on the most widely used summary measures and refer you to the Further Reading section for the more exotic ones (and, believe us, there are plenty of them!). Note that, in what follows, we will be solely concerned with **sample statistics**, that is with measures of central location, variability, etc., based on data obtained from a sample; issues relating to corresponding **population parameters** will be considered in detail in Chapter 9. Moreover, for reasons already explained in Chapter 7, only summary measures based on original (i.e. ungrouped) data will be discussed; methods of calculating equivalent statistics based on data in the form of grouped frequency distributions can be found in the Further Reading section (but read Hint 7.2 again in Chapter 7).

sample statistics

population parameters

When looking at specific measures of average, dispersion, etc., it is useful to bear in mind some general criteria for evaluating alternative measures. While the answer to the question 'What makes a good descriptive statistic?' is just as difficult to give as to 'What makes a good husband or wife?', Table 8.1 lists a number of desirable properties of summary measures (the role of these properties becomes clearer as you fight your way through this and the next chapter).

- A summary measure should be *appropriate for the level of measurement* of the variable in question; e.g. if it involves summation or averaging of values, it should be only applied to metric data.
- It should ideally be *based on all the observations*, i.e. its calculation should make use of the entire set of individual values.
- It should be *simple to comprehend, easy to calculate and expressed in algebraic terms.*
- It should be *unique rather than multivalued*, i.e. given a set of data it should take on *one* value only.
- It should be *resistant to outliers*, i.e. not severely affected by extreme values.
- It should ideally be *equal to actual-data values* and should be *expressed in the original measurement units*, i.e. those of the variable in question.
- It should be *consistent*, i.e. the sample statistic in question should approximate the true value of the corresponding population parameter as the sample size approaches infinity.
- It should be *unbiased*, i.e. if several samples are taken and the sample statistic in question is calculated, then the average of these sample statistics should equal the population parameter.
- It should be *efficient*, i.e. the statistic in question should exhibit minimum variation across different samples.

Table 8.1 Criteria for evaluating summary measures

Measuring Central Location

Our main concern here is with what we may call 'typical' or 'average', that is with computing a *single* value which is in some way representative of the entire set of observations for the variable concerned. Indeed, 'an average is a central reference value which is usually close to the point of greatest concentration of the measurements and may in some sense be thought to typify the whole set'.[1] We shall now look at the three most important location measures, namely the mode, the median and the mean.

The Mode

Undoubtedly, the simplest measure of central location is the **mode**, which is defined as the most frequently occurring value. As it is the value corresponding to a frequency and does not involve any algebraic manipulation of the individual observations, the mode can be applied to any variable irrespective of its level of measurement. However, the mode is particularly useful for summarizing nominal data, as other measures of central location (to be discussed shortly) assume at least ordinal-level data. Note that the mode is not itself a frequency but a value associated with the highest frequency; this is often an issue of confusion, so be careful.

Table 8.2 shows both the ease with which the mode can be determined and the limitations of the mode as a measure of average; the example shown relates to a variable indicating the favourite colour for socks among a small sample of casino owners.

In panel A, the mode is 4, reflecting a clear preference for Shocking Pink socks among casino owners (these guys have taste, don't you doubt it!). Here the value of 4 occurs eight times, much more frequently than any other value, so no problems. Things get a bit more complex in panel B, where all the values occur equally frequently. For all sock colours, the corresponding frequency is four, so identifying *a* modal value is not possible (incidentally, a distribution such as the one in panel B, in which all values occur with equal frequency is known as a **uniform distribution**). In

mode

WARNING 8.1

The mode is an actual value and *not* a frequency of occurrence.

uniform distribution

Table 8.2   The mode in
different situations

Variable	Frequency	Modes
A. Favourite sock colour		
Virgin White = 1	4	
Dawn Scarlet = 2	3	
Sickly Green = 3	3	4
Shocking Pink = 4	8	
Deathly Black = 5	2	
B. Favourite sock colour		
Virgin White = 1	4	
Dawn Scarlet = 2	4	
Sickly Green = 3	4	Uniform distribution
Shocking Pink = 4	4	
Deathly Black = 5	4	
C. Favourite sock colour		
Virgin White = 1	2	
Dawn Scarlet = 2	8	
Sickly Green = 3	1	2 and 4
Shocking Pink = 4	8	
Deathly Black = 5	1	

panel C, we have a different problem as eight casino owners prefer Dawn Scarlet socks and another eight Shocking Pink socks (all other colours are preferred by fewer people). Thus this distribution has two modes (2 and 4) and is commonly referred to as a **bimodal distribution**; other than reporting both modes, there is little one can do here. Clearly, for distributions with two or more modes (i.e. for **multimodal distributions**), it would be misleading to talk about the mode as a measure of 'central' location. The possibility of having multiple modes also exposes another disadvantage of this measure, namely its instability under sampling; for example, taking several samples from a bimodal distribution with population modes M_1 and M_2 is likely to result in some samples having M_1 as their mode and other samples M_2 (thus the mode would fluctuate substantially from sample to sample, which is a pain in the neck to say the least).

bimodal distribution

multimodal distribution

WARNING 8.2

The mode can be considered as a measure of central location *only* for distributions that taper off systematically towards their extremes.

Note that with metric (i.e. interval and ratio) data, it is sometimes possible to have two values with 'the highest' and equally occurring frequencies and still be able to calculate a single mode (and all this without violating Warning 8.2). Consider the following set of values, indicating the number of socks bought by our 20 casino owners in the past week:

8, 9, 10, 10, 11, 11, 11, 12, 12, 13, 13, 13, 13, 14, 14, 14, 14, 15, 15, 16

Here, the values 13 and 14 both occur with a frequency of four, which is greater than the frequency of occurrence of the remaining values. As the two 'modal' values are adjacent, the mode may be arbitrarily taken to be the arithmetic average of the two values (i.e. in this case $(13+14)/2 = 13.5$). It goes without saying that this 'averaging' procedure is only legitimate if *adjacent* values of a *metric* variable are involved. Where two non-adjacent values occur with equal frequencies which are higher than the remaining frequencies, then the distribution must be treated as bimodal and Warning 8.2 applies. Similarly, with non-metric (nominal and ordinal) data, averaging of values (whether adjacent or not) is *not* legitimate for reasons explained in detail in Chapter 3 (think about the outcome of averaging sock colours and you will immediately see why).

Note that, even with **unimodal distributions**, the mode may not be very informative of the structure of the data because the 'most frequently' occurring value may not occur very often. For example, in the following set of values, the mode is 8, but you could not possibly designate it as the *typical* value (with a straight face, that is!).

unimodal distribution

1, 2, 3, 6, 7, 8, 8, 9, 12, 24, 25, 26, 38, 44, 52, 53, 54, 112, 313, 414

All in all, the mode does have its share of problems (and who doesn't after all?) and would not score highly on most of the evaluation criteria listed in Table 8.1. However, as it is the only measure of average that is available for nominal data, we should not complain too loudly. Nevertheless, use it with care.

The Median

The **median** is another measure of central location and is defined as the value above and below which one half of the observations fall. In other words, the median is the value of the 'middle' case when all individual observations have been arranged in rank order. Needless to say, calculating a median for nominal data is totally meaningless as nominal values cannot be ordered (see Chapter 3 if you are not clear on this point).

median

When the number of observations (i.e. sample size) is odd, one value is always in the middle and this is the median; for example, consider the following ordered set of seven values:

5, 5, 4, 3, 2, 2, 1

The median is the fourth value, i.e. 3. When the number of observations is even, then there are two middle values and it is customary to think about the median as being halfway between the two middle values. Consider, for example, the following ordered set of eight values:

5, 5, 5, 4, 3, 2, 2, 1

Here the median lies between the fourth and fifth values, i.e. between 4 and 3. What happens now depends on the level of measurement of the variable in question. If the variable concerned is interval or ratio, then it is usual to take the arithmetic average of the two middle values as the median, in this case $(4 + 3)/2 = 3.5$. On the other hand, if the variable concerned is only ordinal, averaging of values is not legitimate and we can only say that the median lies between the fourth and fifth values (i.e. between 3 and 4) in the ordered set of observations (although, in practice, 'illegal' averaging of the two middle values with ordinal data is commonplace).

WARNING 8.3

Make sure that you have *at least* ordinal-level data before you even *think* about calculating a median!

WARNING 8.4

Do not confuse the *position* of the median with the *value* of the median. The former indicates *where* the median is located in relation to the individual (ordered) observations, while the latter indicates *what* the median value is.

In general, the location of the median, given an ordered set of values, is found by computing $(n+1)/2$, where n = sample size. Thus, in our first example above, the median is positioned at the $(7+1)/2 = 4$th value, while in our second example, the median is located at $(8+1)/2 = 4.5$th value (i.e. midway between the 4th and 5th values). Note that the formula $(n+1)/2$ does not give the *value* of the median, only its *location*.

Sometimes people have difficulties in determining the median from a frequency distribution, as a result of confusing the location and actual value of the median. Table 8.3 recasts our two examples above in frequency distribution form (the variable involved being a 5-point ordinal scale describing the perceived quality of food at *The Dodgy Chicken*, a notorious local takeaway). In panel A, you should immediately see that the median rating is 'passable' (the value of the fourth case), while in panel B, the median rating is between 'passable' and 'good' (the values of the fourth and fifth cases, respectively). If you cannot see these medians *at a glance*, then please go over the examples again. Then check that you've *really* got the hang of it, by going to panels C and D and determining the medians for the data shown.

Table 8.3   The median in different situations

Variable	Frequency	Median
A. Quality rating of Dodgy Chicken		
Excellent = 5	2	
Good = 4	1	
Passable = 3	1	3
Poor = 2	2	
Atrocious = 1	1	
B. Quality rating of Dodgy Chicken		
Excellent = 5	3	
Good = 4	1	
Passable = 3	1	Between 3 and 4
Poor = 2	2	
Atrocious = 1	1	
C. Quality rating of Dodgy Chicken		
Excellent = 5	1	
Good = 4	1	
Passable = 3	3	?[a]
Poor = 2	1	
Atrocious = 1	1	
B Quality rating of Dodgy Chicken		
Excellent = 5	2	
Good = 4	2	
Passable = 3	2	?[a]
Poor = 2	1	
Atrocious = 1	1	

?[a] = *you* tell us!

The median has two advantages as a measure of central location. First, unlike the mode, it is based on the whole distribution of a variable (as one needs to rank the observations in order to locate the median). Second, the median is not affected by extreme values; this makes it an attractive measure even for metric data, particularly when **outliers** (i.e. extreme or atypical values) are involved. For example, the medians for the following three sets of data (A, B and C) are identical (we leave it up to you to decide what the median value is!).

outliers

A: 5, 6, 9, 10, 15, 16, 20
B: 5, 6, 9, 10, 15, 16, 200
C: 1, 1, 1, 10, 20, 400, 6000

The Mean

The **mean** is your familiar arithmetic average and is defined as the sum of a set of values divided by their number; as its computation involves algebraic manipulation of the individual data values, the mean is an appropriate measure of central location for metric data only. As an example, consider the following data indicating the weight in kilograms of six Sumo wrestlers (after they had a light breakfast):

mean

175, 233.7, 199.6, 304.7, 254.9, 388

The mean weight of these healthy lads comes to

(175+233.7+199.6+304.7+254.9+388)/6 = 1555.9/6 = 259.3 kg

In general, given n observations with values $X_1, X_2, X_3, \ldots, X_n$, the mean is found by applying the following ridiculously simple formula:

$$\bar{x} = \frac{X_1 + X_2 + X_3 + \ldots + X_n}{n} = \frac{\sum X_i}{n}$$

The mean has a number of properties worth noting. First, it makes full use of the data available, in that its calculation is based on all the individual data values. The flip side of this is that it can be greatly affected by one or two extreme values (outliers); as such the mean is said to be a **non-resistant measure**. For example, while the median in both sets of data below is 13, the mean is 14.1 in A and 24.1 in B.

non-resistant measure

A: 8, 9, 12, 13, 18, 19, 20
B: 8, 9, 12, 13, 18, 19, 90

Another rather annoying characteristic of the mean is that it can take fractional values, even when the variable involved is discrete (and, thus, only takes integer values – see Chapter 7). For example, if data set A above represented the number of children of seven families in Rabbitsville, Pennsylvania, the average number of children comes to 14.1 – which raises the interesting biological question as to what 0.1 of a child actually means! However, it is only fair to say that this 'is a problem of how fractional values of the mean should be interpreted, rather than a limitation of the mean itself'.[2]

A related issue is that the mean can take a value which does not reflect any of the individual values of the variable in question; for example, if the Domingo family has 2 cars, the Pavarotti family 18 and the Caruso family 16, the mean number of cars comes to (2+18+16)/3=12, a value which

is not representative of any of the families.

But not all is bad with the mean – quite the contrary. First, the mean is unique in the sense that the total sum of (signed) deviations around it is zero (i.e. $\sum(X_i-\bar{x}) = 0$). Second, and related to the first point, the sum of negative deviations from the mean is always equal to the sum of positive deviations from the mean (use data sets A and B above to verify that this is so). This gives the mean a special interpretation as a balance point (or 'centre of gravity') for the distribution of individual values, as negative deviations are exactly offset by positive deviations. Third, the sum of **squared deviations** around the mean, $\sum(X_i-\bar{x})^2$, is smaller than the sum of squared deviations around any other value; this property is used in the calculation of measures of dispersion as we will shortly see. Moreover, without wishing to go into details, you should note that this **least squares** criterion underpins many statistical techniques. Finally, the mean is much more stable than the median or mode over repeated sampling (it exhibits less variation from sample to sample compared to other measures of central location).

Having looked at the characteristics of three different central location measures, it is useful to briefly consider them in conjunction with each other. This is both because no individual measure is ideal on all criteria listed earlier in Table 8.1 and because different measures can furnish complementary perspectives on the same set of data (assuming, of course, that the level of measurement for the variable concerned allows the calculation of all three measures).

Take a look at the fictitious distribution shown in Figure 8.5. Its mode is the point on the horizontal axis which corresponds to the 'peak' of the curve. Its median is a point on the horizontal axis where the ordinate divides the area under the curve into two equal parts; half the area of the curve falls to the right of the ordinate at the median and half to the left. Lastly, its mean is also a point on the horizontal axis, reflecting the centre of gravity of this distribution; if we were to cut out this distribution and try to balance it on the edge of a ruler, the point of balance would be the mean (now, don't get carried away and deface your book just to try this out!).

If you are lucky enough to have a unimodal, symmetrical distribution, then the mode, the median and the mean will all coincide at the centre of the distribution (see Figure 8.6(a)). In contrast, with a positively skewed distribution, the mean will have the largest value and the mode the smallest value, with the median in between (see Figure 8.6(b)). With a negatively skewed distribution, on the other hand, the mode has the largest value and the mean the smallest; again, the median will be between the mean and mode (see Figure 8.6(c)).

squared deviations

least squares

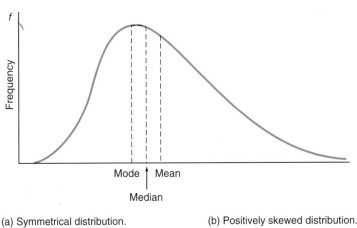

Figure 8.5 The mean, median and mode in a distribution.

(a) Symmetrical distribution.   (b) Positively skewed distribution.

Figure 8.6 Central location measures in different distributions.

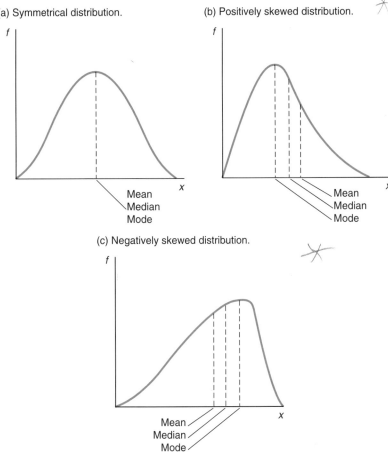

(c) Negatively skewed distribution.

The fact that the *relative* magnitudes of the mode, median and mean differ according to the shape of the distribution suggests that we can use them to get some rough indication of skewness (there are also formal indices to assess skewness as we will see later on). In this context, we usually rely on the relationship between the median and the mean alone, as the mode is the least reliable of all measures of average (particularly in small samples). Thus, if the mean is larger than the median the distri-

HINT 8.1

Use several central location measures in conjunction with one another to provide a comprehensive picture of your data.

bution is positively skewed; if the mean is smaller than the median the distribution is negatively skewed.

Okay, so what measure of average should be used when? First, make sure that you do have a choice, i.e. that your level of measurement allows you to pick among the three measures! Assuming this is so, then do not use the mean (at least not on its own), if your distribution is markedly skewed, particularly when one or more very extreme values are at one side of the distribution; opt for the median instead. With roughly symmetrical distributions, the mean is probably the best measure to use, although (to be on the safe side) it would not harm to report the median as well. Finally, as far as the poor mode is concerned, it should not be reported on its own but only used to 'spice up' a picture painted by the median and/or mean (unless, of course, you are dealing with a nominal variable and the mode is your only choice).

At this point, you should note that there are many more central location measures available with such colourful names as the 'midextreme', the 'midhinge' (nothing to do with door support), the 'trimmed mean', the 'windsorized mean', the 'geometric mean' and the 'harmonic mean' to name but a few. The references in the Further Reading section should be seriously consulted by those of you who wish to impress members of the opposite sex by subtly displaying your knowledge of different measures of average.

Measuring Variability

While measures of central location identify 'average' values in a set of data, they tell us absolutely nothing about the extent to which the individual values are similar or different from one another. However, in many cases, information concerning variability among a set of values is more valuable than information about the 'average'. Consider the following two data sets, indicating the number of lost working days (due to hangovers) by a sample of seven French wine tasters:

October 1995:   4, 5, 5, 5, 5, 5, 6
November 1995:   0, 1, 3, 5, 5, 9, 12

In both months, mean = median = mode = 5 days (i.e. all their averages are the same). Despite this, however, we can see that the pattern of days lost in November is substantially different compared to that in October. In October, all wine tasters lost between 4 and 6 days due to hangovers (an unfortunate professional hazard). In November, the number of days lost ranged from none (by the 'employee of the month') to 12 days (no doubt by the 'bad employee of the month'). Clearly, the extent of variation in November is much greater than in October.

Measures of variability provide a complementary function to central location measures by summarizing the degree of dispersion (or 'scatter' or 'spread') in a variable. They all equal zero when there is no variation and increase in value with greater dispersion in the data. We will look at the following variability measures in this section: the index of diversity for nominal variables; the range and interquartile range for ordinal variables; and the variance and standard deviation for interval and ratio variables.

The Index of Diversity

The **index of diversity** (D) is a measure of variability for nominal data and is based on the frequencies associated with the variable in question. It involves (a) determining the relative frequencies (i.e. proportions) in each category of the variable, (b) squaring them, (c) summing the squares, and (d) subtracting the resulting sum from 1 (phew!). For example, using the data on sock colour preferences in Table 8.2 (panel A), gives us

$$D = 1 - [(4/20)^2 + (3/20)^2 + (3/20)^2 + (8/20)^2 + (2/20)^2] = 1 - 0.255 = 0.745$$

This measure shows the degree of *concentration* of the cases in a few large categories (as squaring relative frequencies affects the large frequencies much more than smaller ones) and tends to zero if almost all cases fall in the same category. It is at a maximum when each category occurs just once; however, its maximum value depends partly on the number of categories: thus, D cannot be used to compare nominal variables with different numbers of categories. However, given a number of categories c, one can calculate the maximum value of the index, which is $(c - 1)/c$, and then compare it to the actual value obtained from the data. In our example, the maximum diversity is $(5 - 1)/5 = 0.80$ (since there are five categories in our variable – see Table 8.2). Given that the actual value of D is 0.745 for our data, we can conclude that there is quite a bit of variation among casino owners in terms of sock colour preferences.

Before we move on, you should note that measures of variability for nominal data are not used very often (nobody really knows why!). Indeed, most standard texts in statistics do not even mention them and, what is infinitely worse, many computer packages do not compute them! Yes, that's right – *you* have to do the work! While you can find out more about them by following the excellent sources in the Further Reading section, please do not hold us responsible for bumping into such weird measures as the dreaded 'standardized entropy index' or the much-feared 'fragmentation statistic'.

The Range and Interquartile Range

The range and interquartile range are measures of variability particularly useful for ordinal variables. The **range** is simply defined as the difference between the highest and lowest values in the data; when there is no variation in the variable, the range is zero. While it is extremely easy to calculate, the range does not use all the information in the data and, thus, is very much affected by extreme values. For example, the range for both data sets below comes to 19, although arguably the fluctuation in values in A is greater than in B.

index of diversity

range

A: 1, 2, 5, 9, 11, 20
B: 1, 1, 1, 1, 1, 20

For large samples, the range is an unstable measure of variability as its fluctuation from sample to sample increases as the sample size increases. Further, there is a higher chance of obtaining extreme values with large samples than with small samples; what this implies is that, for the same variable, ranges calculated on different sample sizes are not directly comparable.

To deal with the above problems, it makes sense to opt for a modified range, which is established by eliminating a certain percentage of the extreme values at the two ends of the distribution (thus providing a more stable and reliable measure of variability). One such modified range (there are others) is the **interquartile range** (yes, it does sound like a testing ground for ballistic missiles!). It is calculated by (a) finding the values corresponding to the 75th and 25th percentiles (i.e. the upper and lower quartiles, Q_3, Q_1), and (b) subtracting the latter from the former (i.e. forming the difference $Q_3 - Q_1$); any decent computer package should be able to provide you with Q_3 and Q_1 (if your package cannot, then you can always go to Chapter 7 and follow our superb guidelines for finding percentiles). As a quick example, in the following data set the range comes to 39, while the interquartile range comes to 18 (and if you cannot tell how we have arrived at these figures, you should brush up on your percentiles).

interquartile range

  5, 10, 25, 28, 44

There are several other measures of variability for ordinal data, and, if you will not be satisfied until you learn everything there is to know about the 'coefficient of quartile variation' or 'Leik's ordinal consensus measure', the Further Reading section is eagerly awaiting you.

The Variance and Standard Deviation

These are by far the most widely used and highly regarded measures of variability – you cannot open any statistical text and find no reference to the variance and standard deviation (it's like opening *Cosmopolitan* magazine and finding no reference to sex!). They are widely used not only as descriptive measures but also in connection with inferential statistics (see Chapters 9 and 10). So what are these apparently amazing dispersion measures?

variance

The **variance** involves (a) subtracting the mean from each individual value, i.e. forming all deviations from the mean; (b) squaring the deviations, (c) summing them; and (d) taking their average. Thus, using the horrible notation introduced previously, the variance is defined as

$$\frac{\sum (X_i - \bar{x})^2}{n}$$

where X_i are the individual values, $\bar{x}$ is the mean and n is the number of observations. Unfortunately, when calculated using sample data, the above

formula provides a **biased estimate** of the population variance; specifi-   biased estimate
cally, it shows a systematic tendency to underestimate the population
variance (why this happens need not concern you here). As this is clearly
undesirable (see Table 8.1), we use the following slight modification to calculate the sample variance s^2; this gives an **unbiased estimate** of the   unbiased estimate
population variance, so everybody is happy:

$$s^2 = \frac{\sum(X_i - \bar{x})^2}{n-1}$$

As an example, let us calculate the variance for the following small set
of data, indicating the number of domestic cats reported lost in 12 police
stations in Wigan, Lancashire:

37, 59, 71, 75, 78, 78, 81, 86, 88, 92, 95, 96

The mean number of cats reported lost is

$(37+59+71+75+78+78+81+86+88+92+95+96)/12 = 78$,

so the variance is

$[(37–78)^2+(59–78)^2+(71–78)^2+\cdots+(96–78)^2]/(12–1) = 3082/11 = 280$

One clear advantage of the variance is that it takes into account each
and every piece of data available, as all the individual values are used in its
calculation. Second, it expresses variation in individual values in relation
to a measure of central location, namely the mean; when the values are
scattered widely about the mean the variance is large and when the values
are concentrated near the mean, the variance is small. Third, the variance
emphasizes large deviations from the mean as these, when squared,
increase much more than small deviations; however, while large deviations from the mean indicate greater variability, squaring the deviations
implies that extreme values can have a disproportionate impact, making
the variance a non-resistant statistic.

Also on the down side, the variance is unfortunately expressed in
squared units of the variable in question. In the above example, the variance of 280 does not refer to 'number of cats' but 'number of cats squared'!
As this makes interpretation rather messy, it is usual to report the square
root of the variance: this is the famous **standard deviation**. Thus, in our   standard deviation
example, the variability across police stations would be described in stan-
dard deviation terms as being 16.7 cats ($\sqrt{280}$). In general, the sample
standard deviation s is defined as

$$s = \sqrt{s^2} = \sqrt{\frac{\sum(X_i - \bar{x})^2}{n-1}}$$

where all the symbols are as before.

While describing the dispersion of a variable using the standard deviation is preferable to using the variance because the original measurement

units are retained, it may still be difficult to interpret standard deviation values directly. For example, is a standard deviation of £100 large or small? It may be considered large if we are measuring the weekly amount spent on sweets but small if we are looking at the incomes of statistics professors. In cases such as this, it may be more helpful to look at the standard deviation in relation to the mean for the variables in question; in this way, we can make comparisons across different variables. A measure that accomplishes this beautifully is the **coefficient of variation** *CV*, which is defined as the standard deviation divided by the mean, or

coefficient of variation

$$CV = \frac{s}{\bar{x}}$$

Thus, if the mean weekly amount spent on sweets is £125 and the mean weekly professorial salary for statisticians is £1300, then the respective coefficients of variation come to 100/125 = 0.80 and 100/1300 = 0.077, respectively; thus the amounts spent on sweets are relatively much more variable than the incomes. Note that the coefficient of variation is independent of the unit of measurement (i.e. it is 'unitless'). This is because both the standard deviation and the mean are in the same units, which cancel out when their ratio is formed.

The standard deviation can also be used with the mean to compute what are known as **standard scores** (or 'z-scores'). Sometimes the information provided by a raw value has little meaning unless seen in relation to the other values. For example if you score 63% on a *Creative Accounting* exam, whether you did really well or badly would depend on how the rest of the class did. If the mean grade was 12% then you are flying (and all your colleagues will hate you!). If, on the other hand, the mean grade was 92%, there is little cause for celebration. A standard score allows us to place an individual case in context; it shows us how many standard deviations above or below the mean a certain observation lies. To calculate a standard score from raw scores (what we eloquently call a process of **standardization**), we (a) subtract the mean from the value of the individual observation (i.e. find the deviation from the mean), and (b) divide this difference by the standard deviation; thus

standard score

standardization

$$z_i = \frac{X_i - \bar{x}}{s}$$

Standard scores always have a mean of zero and a standard deviation of unity; however, the shape of the distribution of standard scores is identical to the distribution of raw values (this is important as we will see in a later section). A positive standard score indicates that the observation in question has a value greater than the mean, while a negative score indicates the opposite. Thus, if you converted your *Creative Accounting* grade to a standard score and found z = 1.25, you could brag to your friends that you scored 1.25 deviations above the mean grade awarded to the class as a whole; conversely, if you were unlucky enough to find that z = −1.25, you can be ready to change the subject if a conversation concerning grades ever comes up.

Another important use of standard scores is that they enable comparisons to be drawn between different distributions. For example, if you wanted to compare your performance in the *Creative Accounting* exam with that in the *Unfair Taxation* exam, standard scores should do the trick. If you scored 63% in the former and 70% in the latter, does that mean that *Unfair Taxation* is your better subject? Not necessarily – it depends on the mean score and the standard deviation of scores in each of the two exams. If the mean grade for the class for *Creative Accounting* was 52% with a standard deviation of 8%, while the respective figures for *Unfair Taxation* were 75% and 4%, your standard scores would be

Creative Accounting: (63–52)/8 = 1.375
Unfair Taxation:　　(70–75)/4 = –1.25

Thus, in relation to the other students taking the two exams, you did much better in *Creative Accounting* than in *Unfair Taxation*.

Standard scores are also helpful if one wants to combine several variables to create a **composite scale** (or 'index'). If the variables in question have very different variances or are measured in different units then, obviously, combining (e.g. adding) their raw scores is inappropriate. In such cases, the variables should first be standardized and the standard scores used to form the desired index.

composite scale

There is yet another important use of standard scores; we will see what that is after a brief excursion into the wonderful world of skewness and kurtosis.

Measuring Skewness and Kurtosis

Although, for most practical purposes, one can get a reasonable idea of skewness and kurtosis by looking at a histogram or frequency polygon and by examining the relative magnitudes of the mean and median (see earlier), it is sometimes desirable to compute some formal measures. A measure of skewness or kurtosis will be close to zero if the distribution of values is symmetric and mesokurtic (i.e. like the nice bell-shaped curve of the normal distribution discussed later in this chapter). Note that with sample data it is extremely unlikely that *exactly* zero values will be obtained, because of sampling fluctuations. Positive values for skewness indicate a positive skew and positive values for kurtosis reflect a more leptokurtic distribution.

As the formulae for skewness and kurtosis are rather scary, and given that decent computer packages can provide a **coefficient of skewness** and a **coefficient of kurtosis** upon request, we will restrain ourselves and will not reproduce the formulae here. Instead we will present you with an example of *SPSS/PC+* generated computer output showing a variety of summary measures (Table 8.4); the variable in question shows the age (in years) of a sample of 80 snake charmers.

coefficient of skewness
coefficient of kurtosis

We can see that the mean and median ages are practically the same, but the modal age is lower; this suggests that the distribution may be somewhat positively skewed. This is indeed the case, as indicated by the positive

Table 8.4 An example of computer-generated descriptive output

AGE	Snake charmer age (in years)					
Mean	40.890	Median	41.000	Mode	30.000	
Std dev	9.379	Variance	87.960	Kurtosis	−0.279	
Skewness	0.359	Range	41.000			
Valid cases	73	Missing cases	7			

coefficient for skewness; moreover, the negative coefficient of kurtosis suggests that the distribution of ages is rather platykurtic.

Chebychev's Theorem and the Normal Distribution

After this brief encounter with skewness and its mate, kurtosis, we can now return to more important issues. The first thing to consider is a wonderful theorem developed by a clever Russian, named Chebychev, well over 100 years ago. This enables us to calculate for *any* set of data, the *minimum* proportion of values that can be expected to lie within a specified number of standard deviations from the mean. According to this theorem, at least 75% of the individual values in a set of data can be expected to fall within two standard deviations from the mean, at least 88.9% will fall within three standard deviations from the mean, and at least 96% of all values will be within five standard deviations from the mean. More generally, **Chebychev's theorem** tells us that given a set of observations, the probability is at least $(1-1/k^2)$ that an individual observation will take a value within k standard deviations from the mean (where $k > 1$).

Chebychev's theorem

To illustrate the application of the above theorem, let us use the snake charmer data in Table 8.4. We know that the mean age (rounded) is 41 and the standard deviation about 9 years (also rounded). If we set $k = 2$, then the probability is at least $(1-1/2^2) = 0.75$ that the age of a snake charmer will be within two standard deviations (i.e. within $2 \times 9 = 18$ years) from the mean (which is 41). In other words, we can expect that at least 75% of the snake charmers will be between 41−18 = 23 and 41+18 = 59 years old. Isn't this great? We can, of course, select any value for k and compute the relevant intervals for our variable. Note that, given the same mean and a specified value for k, the intervals will be narrower the lower the standard deviation (try the example as above but using a standard deviation of 6 years and you will see what we mean). This is because, for a set of data with a small standard deviation, a larger proportion of the values will be concentrated near the mean than for a data set with a large standard deviation.

The key point to take home from all this is that by using the standard deviation in conjunction with the mean we are able to make some *probabilistic* statements (i.e. 'informed guesses') about the position of individual values relative to the mean. Note that we are able to make such statements without making any assumptions about the specific form or shape of the distribution; indeed, Chebychev's theorem can be applied to any distribution of values (and, indeed, this is one of its advantages). The question now is: Can we do any better if we know something about the form of the distribution concerned?

The answer is yes. Consider the distribution shown in Figure 8.7. This is a symmetric, bell-shaped, mesokurtic (neither flat nor peaked) distribution known as the **normal curve** (yes, this is the normal distribution – say hi to it!). Its mean, median and mode all coincide and each half of the distribution is a mirror image of the other half; moreover, the tails of the curve tend towards but never actually touch the horizontal axis (i.e. this 'you can reach but never touch' quality, makes the normal distribution *asymptotic*). A key characteristic of this distribution is that approximately 68% of the individual values fall within one standard deviation from the mean, approximately 95% of individual values fall within two standard deviations from the mean and approximately 99.7% of all values within three standard deviations from the mean.

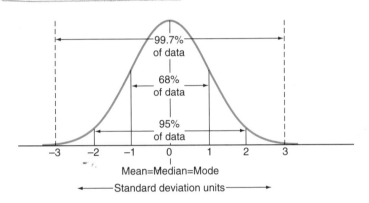

normal curve

Figure 8.7   The normal distribution.

Let us now use our snake charmer data and calculate the proportion of ages within two standard deviations from the mean, assuming a normal distribution for the age variable. As before, the mean is 41 and the standard deviation 9 years. Thus the desired interval is defined by 41–2(9) = 23 and 41+2(9) = 59 which, assuming a normal distribution, indicates that approximately 95% of the snake charmers are between 23 and 59 years old. If we compare this interval with that established using Chebychev's theorem, we see that, while they are identical in size, we now include 95% as opposed to 'at least 75%' of the individual cases (see above). In other words, whereas before we could only say that there is at least a 0.75 probability that an individual value will lie within two standard deviations from the mean, we are now able to attach a 0.95 probability to the same estimate. We are able to do this because we know the *form* of the distribution involved (in this case the normal) and we can use this information to our advantage in order to make more precise probabilistic statements (i.e. more informed guesses). Aren't you impressed?

The normal distribution is just one of several distributions with known properties. Other well-known distributions include the **binomial distribution**, the **hypergeometric distribution**, the **Poisson distribution**, the **chi-square distribution**, the ***t*-distribution** and the ***F*-distribution** to name but a few. In each of these distributions, the probabilities associated with all possible values are known, and for this reason they are often

binomial distribution
hypergeometric distribution
Poisson distribution
chi-square distribution
***t*-distribution**
***F*-distribution**

probability distributions

referred to collectively as **probability distributions**. While a detailed discussion of different probability distributions is way beyond the scope of this text (we do not want to make enemies of you!), there are some basic things you ought to know.

First, we can distinguish between probability distributions which have been developed for discrete variables and those for continuous variables (see Chapter 7, if you have forgotten the difference between the two). The former are known as **discrete probability distributions** and include the binomial, hypergeometric and Poisson distributions. The latter are referred to as **continuous probability distributions** and include the normal, t-, and F-distributions.

discrete probability distributions
continuous probability distributions

Second, some of these distributions are symmetrical about their mean (e.g. the normal and t-distributions) whereas others are not (e.g. the F- and chi-square distributions). This is an important distinguishing characteristic, the implications of which will become evident in Chapter 11.

Third, the specific form of some probability distributions (e.g. the chi-square distribution) is dependent upon what is known as **degrees of freedom**, which is a rather complex concept to explain in simple terms (so we will not even try!). Suffice it to say that distributions dependent upon the degrees of freedom are really families of distributions (a different distribution is associated with each different number of degrees of freedom).

degrees of freedom

Probability distributions are also known as **theoretical distributions** because they are used as models to study real-life variables that may be approximately distributed as the theoretical distributions. Put differently, we often use theoretical distributions to study **empirical distributions**, i.e. those resulting from actual data. We also use theoretical distributions in our efforts to estimate population parameters (see Chapter 9) and to test hypotheses (see Chapter 10). Thus the 'idealized' distribution types that statisticians have developed over the years are very important for data analysis purposes: they provide the foundations for inferential analysis and enable us to go beyond mere description of our sample data.

theoretical distributions

empirical distributions

By far the most important theoretical distribution is the normal distribution, which is the backbone of much of statistical analysis. There are several reasons for this. A large number of natural phenomena (e.g. height and weight) are approximately normally distributed; errors of measurement and prediction errors are also usually assumed to be normally distributed; certain sample statistics such as the sample mean are also approximately normally distributed, given a reasonable sample size; a number of other probability distributions (e.g. binomial and Poisson) can be effectively approximated by the normal distribution; the normal distribution is easy to work with because of its symmetrical nature and the fact that it is completely described by its mean and standard deviation; the list goes on.

One particular version of the normal distribution deserves special attention. This is the **standard normal distribution** and is based on the concept of standard scores (see earlier). Also sometimes referred to as the 'unit normal distribution', this is the distribution of normally distributed standard scores. Thus, the standard normal distribution is a normal distribution with a mean of 0 and a standard deviation of 1. Any normal distribution with a given mean and a given standard deviation can be

standard normal distribution

statistical tables

translated into the standard normal distribution, simply by converting the raw scores into standard scores (see earlier). A direct implication of this is that we can use the standard normal distribution as a general model and generate **statistical tables** showing the proportion of cases falling within a certain distance from the mean. In this context, the area under any normal curve is proportional to the frequency of values, so the proportion of values below (above) the mean is represented by the area under the curve that lies below (above) the mean. To determine this proportion (without the help of statistical tables), one would have to use integral calculus – which is as pleasurable an experience as eating live eels (if you have ever seen the mathematical formula describing the normal distribution, you'll know what we mean!). What's even worse, one would have to engage in the dubious pleasure of integration for all possible distances from the mean (in order to find the proportion of cases falling a certain distance above/below the mean or, which amounts to the same thing, to find the probability that a case will fall within a certain distance from the mean). Naturally, this process would have to be repeated every time a different distribution was under study, as a different mean or standard deviation results in a different normal distribution! Indeed, 'the number of possible combinations of means and standard deviations is infinite, so the number of different normal distributions is also infinite'.[3]

This is where the standard normal distribution comes to our rescue. Luckily, there are published tables for the standard normal distribution (developed by geniuses in integral calculus), showing (a) the values of standard scores, (b) the proportion of the area falling between the mean and a given standard score, (c) the proportion of the area falling beyond a given standard score, and (d) the height of the curve (i.e. the ordinate) for the given standard score. Different statistics texts incorporate different combinations of the above tables in their appendices (see Further Reading section). Note that as the total area under the standard normal curve is equal to 1 (i.e. 100%) and as the curve is symmetric, the tables usually cover only one half of the distribution (to the right of the mean); thus only positive z-scores are covered. Figure 8.8 reproduces the standard normal distribution and the areas under the curve for $z = 0, \pm1, \pm2$, and ±3 (recall from our earlier discussion that a standard score reflects the distance from the mean in standard deviation terms, e.g. $z = 2$, indicates that the raw value lies two standard deviation above the mean. The mean is, of course, at $z = 0$).

Enough of theory. Let us see what the standard normal distribution can *do* for you. Just for the sake of it, we will assume that you have a normal distribution with a mean of 20 and a standard deviation of 5 and you want answers to the following questions (you can ask similar questions with any normal distribution and the specific figures involved are for illustrative purposes only):

1. What is the proportion of scores lying *within* 0.5 standard deviation from the mean?
2. What is the proportion of scores *below* 1.5 standard deviations from the mean or *above* 2.0 standard deviations from the mean?

Figure 8.8   The standard
normal distribution.

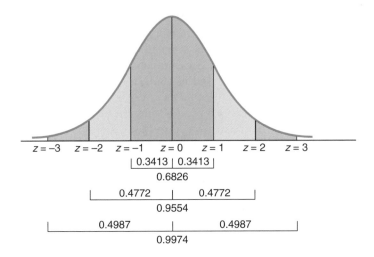

3. What is the proportion of scores *between* 1.0 and 1.5 standard deviations *above* the mean?
4. What is the proportion of scores lying *between* 0.5 standard deviation below the mean and 2.0 standard deviations *above* the mean?
5. Where about in the distribution is a raw score of 8 located?
6. What proportion of scores exceeds a raw score of 22?
7. What is the 70th percentile of this distribution?
8. What is the percentile rank of a raw score of 10?
9. What percentage of raw scores are between 6 and 12?
10. Who shot JR in *Dallas*?

Yes, you can get the answers to all of these questions (well, the first nine anyway!) simply by making use of the tables of the standard normal distribution; here we are using the 'classic' tables of Fisher and Yates (see Further Reading section), but you should be able to replicate the results using any standard normal distribution tables (if not, *you* are doing something wrong, the tables are always right!). Okay, here it goes:

1. This is a real giveaway; all you need to do is go to your statistical tables and look up the area between $z = 0$ (the mean) and $z = -0.50$ and then between $z = 0$ and $z = 0.50$ and add the two areas together; this comes to $0.1915+0.1915 = 0.383$. Thus approximately 38.3% of the observations will lie within ±0.5 standard deviations from the mean. See Figure 8.9(a).
2. This is a bit more tricky, but dead easy too. First, determine the area between $z = 0$ and $z = -1.5$; a quick peek at the tables tells you that this comes to 0.4332. Given that the area to the left of the mean is 0.50 (remember, the total area under the standard normal curve is equal to 1), the area below $z = -1.5$ is $0.50-0.4332 = 0.0668$; make a note of this. Now, do exactly the same for $z = 2.0$. The area between $z = 0$ and $z = 2$ is 0.4772, which means that the area above $z = 2.0$ is $0.50-0.4772 = 0.0228$; make a note of this too. Now, add your two notes together $(0.0668+0.0228 = 0.0896)$ and you have your answer:

approximately 9% of the observations have scores 1.5 standard deviations below the mean or 2 standard deviations above the mean. See Figure 8.9(b).

3. This is truly a piece of cake, since you only need to look at one side of the distribution. First, find the area between $z = 0$ and $z = 1.0$; this comes to 0.3413. Now find the area between $z = 0$ and $z = 1.5$; this comes to 0.4332. As the area between $z = 0$ and $z = 1.0$ is *enclosed* within the area between $z = 0$ and $z = 1.5$, all you need to do is subtract the former from the latter (0.4332–0.3413 = 0.919) and you have your answer. Thus approximately 9.2% of observations have scores between 1.0 and 1.5 standard deviations above the mean. See Figure 8.9(c).

4. This is similar to (3) above (i.e. you are looking for the area between two z-scores), the only difference being that *both* sides of the distribution are involved. Proceed exactly as above and establish the areas between $z = 0$ and $z = -0.50$ and between $z = 0$ and $z = 2.00$; these come to 0.1915 and 0.4772, respectively. As each of these areas falls on a *different* side of the mean, you need to add them to get the total area; this gives 0.1915+0.4772 = 0.6687 which indicates that approximately 66.9% of the observations fall in the interval between 0.5 standard deviation below the mean and 2 standard deviations above the mean. See Figure 8.9(d).

5. You've done this one before. Just transform this raw score to a standard score and then check your standard normal distribution tables. Given that your original distribution has a mean of 20 and a standard deviation of 5, a raw score of 8 corresponds to a standard score of (8–20)/5=–2.40. Thus this raw score lies 2.4 standard deviations below the mean. See Figure 8.9(e).

6. Elementary, my dear Watson. First, you must transform your raw score of 22 to a standard score; this should come to (22–20)/5 = 0.40. Now find the area between $z = 0$ and $z = 0.40$; this is 0.1554. Now since you want to find the area beyond $z = 0.40$, you need to subtract 0.1554 from 0.50 (see also (2) above). This gives 0.50–0.1554 = 0.3446, which indicates that approximately 34.5% of observations have raw scores greater than 22. See Figure 8.9(f).

7. This needs a bit of thought. What you want to do is get the standard score below which 0.70 of the total area lies and then transform it back to a raw score. Now, you know that the half of the standard normal distribution to the left of the mean has an area equal to 0.50; so you need another 0.20 to the right of the mean. So what you must find is a z-score such that the area between it and $z = 0$ is equal to 0.20. Rushing to your standard tables, you should be able to find that the z-score you want has a value of 0.52. Great! Now, translate this back to a raw score. How? What do you mean *how*? If $z = (X_i - \bar{x})/s$ gives you the standard score (see previous section), then if you know z, all you have to do is solve for X_i; specifically

$$X_i = \bar{x} + sz$$

where $\bar{x}$ = mean and s = standard deviation. In your case z = 0.52, $\bar{x}$ = 20 and s = 5. Thus the raw score is 20+0.52(5) = 22.6, and this is the 70th percentile of the distribution in question. See Figure 8.9(g).

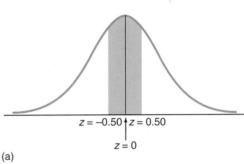

(a)

Figure 8.9   Using the standard normal distribution.

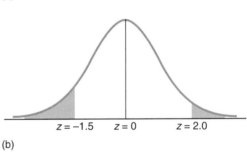

(b)

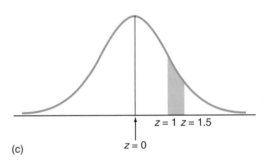

(c)

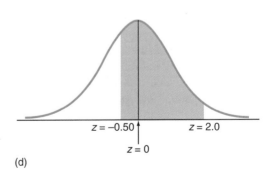

(d)

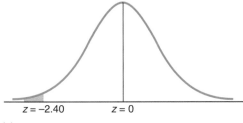

$z = -2.40$ $z = 0$

(e)

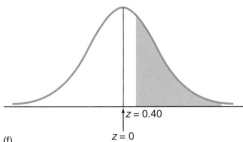

$z = 0.40$

$z = 0$

(f)

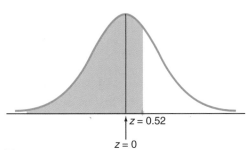

$z = 0.52$

$z = 0$

(g)

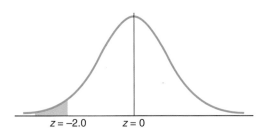

$z = -2.0$ $z = 0$

(h)

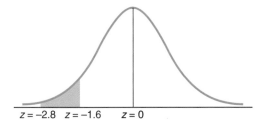

$z = -2.8$ $z = -1.6$ $z = 0$

(i)

8. Compared to (7) above, this is a doddle. First, get your raw score translated into a standard score; you should get $z = -2.00$. Then find the area between $z = 0$ and $z = -2.00$; this is 0.4772. Now subtract this from 0.50 (see also (2) above) and you will get the area below $z = -2.00$ (which is what you want since you are after a percentile rank). Thus the percentile rank of a raw score of 10 is $0.50–0.4772 = 0.0228$, which indicates that approximately 2.3% of the observations have raw scores less than 10. See Figure 8.9(h).

9. This is essentially the same problem as the one in (3) above, the only difference being that you must first transform your raw scores to standard scores. If you do this you will get a standard score of $z = -2.8$ for a raw score of 6 and a standard score of $z = -1.6$ for a raw score of 12. The area between $z = 0$ and $z = -2.8$ is 0.4974 and the area between $z = 0$ and $z = -1.6$ is 0.4452; you now have to subtract the latter from the former and you have the required area, i.e. $0.4974–0.4452 = 0.05222$. Thus approximately 5.2% of the observations have raw scores between 6 and 12. See Figure 8.9(i).

Other probability distributions are used in a similar manner as the standard normal distribution, although their applications are rather more limited. As already mentioned, we will be making much use of different probability distributions in the chapters to come in order to set up confidence intervals (see Chapter 9) and test various kinds of hypotheses (see Chapters 10–13). In fact, directly or indirectly, probability distributions will be with us for the remainder of this book, and thus you should go through the ideas introduced in this section once again to make sure that all is clear (better still, make up some additional examples and play around some more with standard normal tables).

Summary

In this chapter we looked at different summary measures to characterize and compare frequency distributions. First, we showed that different distributions may vary in several respects, namely central location, variability, skewness and kurtosis. Next, we discussed a variety of summary measures for describing these characteristics and highlighted their advantages and limitations. We then examined how central tendency can be looked at in conjunction with variability and learned about standard scores and Chebychev's theorem. Lastly, we introduced the concept of a probability distribution and spent quite a bit of time discussing the normal distribution and the use of statistical tables. We have now concluded our journey into descriptive analysis and are ready to face the challenge of inferential statistics. Enjoy.

Questions and Problems

1. In which ways can frequency distributions differ from one another?

2. What measures of central location could you use for (a) a nominal, (b) an ordinal and (c) an interval variable?

3. What are the potential problems with using the mode as a measure of central location?

4. Why is it important to distinguish between the location of the median and the value of the median?

5. How could you use the mode, median and mean to draw inferences about the skewness of a distribution?

6. Name one measure of variability for each different level of measurement.

7. Under what circumstances would you use the coefficient of variation?

8. What is a standard score? What are its uses?

9. Why is the standard normal distribution so important in data analysis?

10. Did we overdo it by showing you nine different ways of using the standard normal tables?

Notes

1. Ferguson, GA (1981) *Statistical Analysis in Psychology and Education*, p. 48. Tokyo: McGraw-Hill.
2. Weisberg, HF (1992) *Central Tendency and Variability*, p. 30. London: Sage.
3. Hinkle, DE, Wiersnma, W & Jurs, G (1988) *Applied Statistics for the Behavioral Sciences*, p. 90. Boston: Houghton Mifflin.

Further Reading

Fisher, RA & Yates, F (1974) *Statistical Tables for Biological, Agricultural and Medical Research*, 6th edn. London: Longman. Although its entertainment value is equal to emptying a dishwasher, this is *the* collection of statistical tables to have and treasure (however, any set of tables will do).

Huff, D (1973) *How to Lie with Statistics*. London: Pelican. It's short, it's witty, it's brilliant! If you want to learn how to mislead with your data and still get away with it, this book will tell you.

Weisberg, HF (1992) *Central Tendency and Variability*. London: Sage. An excellent little guide to different summary measures (including some very strange ones); we have used it extensively in this chapter, so it must be good!

What about using *estimation* to see what the population looks like?

The Nature of Estimation

In the previous two chapters we saw how one can use different statistical techniques to describe, display and summarize data. While this is an important endeavour in its own right when the focus of the analysis is purely descriptive (see Chapter 6), we often want to go beyond describing our sample data and say something about the population from which our sample was drawn; in other words, we want to make *inferences* about the population on the basis of what we observe in the sample. For example, we may have conducted a survey of 450 randomly selected doctors and found that 63% absolutely hate going to the dentist. Now, although we may have a random sample, can we say that the corresponding percentage in the medical profession *as a whole* is 63%? Could it be higher? Or lower? By how much?

It is questions like these that we try to address every time we engage in estimation. Specifically, estimation can be defined as the process of using a particular **sample statistic** (e.g. mean, standard deviation, proportion) to estimate the corresponding **population parameter**. In this context, 'parameters are fixed values referring to the population and are generally unknown . . . Statistics, on the other hand, vary from one sample to the next . . . In contrast to parameters, the values of statistics are known or can be computed'.[1] Clearly, the notion of the term 'statistics' as used here is different from the everyday usage of the term, i.e. as in published statistics or statistics as a subject area; under the current perspective, a statistic is any quantity or summary measure calculated from sample data, whereas a parameter is any quantity or summary measure calculated from population data.

Let us go back to our example and explore the concept of estimation a little further. In the first instance we may be content to take the sample proportion (0.63) as being our 'best' guess for the population proportion. In doing so, we would be using our sample statistic as a **point estimate** of the relevant population parameter; in general, a point estimate is a *single value* that is obtained from sample data and is used as the 'best guess' of the corresponding population value. Thus, in our example, the point estimate of the population proportion is 0.63. Now, the problem with a point estimate is that, since it is based on sample data, it reflects not only the relevant population parameter but also sampling error. As discussed in

sample statistic
population parameter

point estimate

Chapter 2, sampling error reflects the difference between a result based on a sample and that which would have been obtained if the entire population in question (here all medical doctors) had been studied. Thus, if π represents the (unknown) population proportion and P our sample proportion, then

$$\pi - P = e$$

where e = sampling error.

Three things should become immediately obvious. First, in any one sample, sampling error can be positive or negative; what this means is that a population parameter may be overestimated or underestimated by a sample statistic used as a point estimate (in our example $P = 0.63$ so, depending on whether $\pi > 0.63$, or $\pi < 0.63$, then $e > 0$ or $e < 0$).

Second, the magnitude of sampling error will vary from sample to sample; if we were to take several samples and compute the sample proportions, some of them would be likely to be better point estimates of the population proportion than others (if we took, say, twenty different samples of 450 students and calculated P, could we really expect that P would be *exactly* equal to 0.63 *every single time?*).

Third, the variation in sampling error across samples is solely due to the variation in the sample statistic (here the sample proportion) as the value of the population parameter is always fixed. Although π is unknown, being a population parameter, it cannot vary in value as its computation includes all the elements in the population; in contrast P can vary as it is based only on a part of the population, i.e. those population elements that happen to be included in our particular sample.

If we pull the above observations together, we can see that if we limit ourselves to a point estimate of the population parameter, that estimate will necessarily reflect the specific characteristics of the particular sample concerned. What seems to make more sense is to try to establish a *range* for our estimate by taking into account the likely variation in sampling error. This is exactly what the purpose of *interval* estimation is: to predict that the population parameter in question is somewhere within a given interval either side of a point estimate. In our example, an **interval esti-**
mate would take the form of a statement such as 'the proportion of medical doctors who hate going to the dentist is somewhere between 0.59 and 0.66'.

interval estimate

This is all very well, you may say, but how do we go about constructing such an interval? Given that we only know the value of the sample statistic for our particular sample and given that the population parameter is unknown, we are clearly unable to specify the exact magnitude of sampling error involved. However, if we took repeated samples of the same size and calculated the sample proportions, then we should gain some appreciation of sampling error by observing how these repeated measurements varied from each other; if the sample proportions fluctuated little from sample to sample, then we would gather that the sampling error was less in magnitude than if they fluctuated a lot. Now, imagine that we took *all* possible samples of a given size and calculated the sample proportion. We would eventually end up with a *distribution* of all possible sample

sampling distribution of a proportion

proportions, representing all samples of a given size. We call this distribution the **sampling distribution of a proportion**. In general, the sampling distribution of some statistic (e.g. mean, standard deviation, proportion) can be defined as the distribution of all possible values that can be assumed by this statistic as computed from samples of the same size drawn at random from the same population.

Now, as with all distributions for metric data (and recall from Chapter 3 that dichotomous variables can be treated as such), we could use the standard deviation to measure the variability of this distribution (see Chapter 8). We call this standard deviation the **standard error of the proportion** and denote it with s_p (to distinguish it from the sample standard deviation, s).

standard error of the proportion

Thus the standard error of the proportion is a measure of the fluctuation in proportions from sample to sample and we can, therefore, use it to describe the variation in sampling error. Before we go any further you should be absolutely clear that the standard error of the proportion is (a) a standard deviation of a particular distribution, (b) that this distribution is neither the distribution of individual values in the sample, nor the distribution of values in the population, but (c) the distribution of sample proportions calculated from all samples of a given size selected randomly from the population – what we call the sampling distribution of a proportion. The same applies to any standard error (e.g. the standard error of the mean): a standard error is always the standard deviation of the sampling distribution for the statistic under study.

WARNING 9.1

Do not confuse the standard deviation of the sampling distribution (i.e. the standard error) with either the sample standard deviation or the population standard deviation.

Hang on a second you may shout – what's the point of attempting to construct a sampling distribution so that we can measure the fluctuation in sampling error if we need to take all possible samples of a given size from the population? Wouldn't it be just as quick (or even quicker) to conduct a census, i.e. forget about sampling altogether and cover the whole population instead? After all, if we take all possible samples of a certain size, we should, in one sample or another, include each and every one of the population elements (and thus we might just as well get them all together in one go and be done with it). This is absolutely true if we actually had to go through the procedure outlined above, i.e. if we had to construct a sampling distribution empirically. However, in practice, we do not have to do this because we can use **theoretical sampling distributions**. These are probability distributions with known properties (see Chapter 8) and can be used in conjunction with a single sample to generate estimates of sampling error. Thus there are theoretical sampling distributions for the sample proportion, the sample mean and the variance, to name but a few. Such sampling distributions have been derived mathematically by incredibly bright statisticians (for goodness sake, don't ask us *how!*) but can be taken 'off the shelf' by mere mortals such as ourselves every time we want to estimate a certain population parameter (for example, assuming a normally distributed population, the sampling distribution of the mean is a normal distribution, while the sampling distribution of the variance approximates a chi-square distribution). Thus all we have to do is use the value of our sample statistic together with the standard error of the appropriate sampling distribution and generate an interval estimate as follows:

theoretical sampling distribution

Population parameter = sample statistic $\pm$ k(standard error)

where k = number of desired standard errors for the estimate.

Thus, if we wanted to estimate the population proportion we would (a) calculate our sample proportion, P, based on our sample data, (b) look at the theoretical sampling distribution for sample proportions and calculate its standard error s_p (there are standard formulae for this), and (c) set up an interval by adding/subtracting so many standard errors, k, from the sample proportion; thus our estimate would take the following form:

$$\pi = P \pm ks_p$$

So far so good – but we still have not shown you an actual example of how to get an interval estimate for the dreaded proportion of doctors who despise going to the dentist. What is s_p for the specific sample of 450 doctors under consideration? How small/large should k be? Who will win the pole vault in the next Olympics? Patience, my friends – all will soon be revealed.

Setting Confidence Intervals

Consider the following question: Is an interval estimate that places the population proportion between 0.60 and 0.70 necessarily better than another estimate which places the population proportion between 0.40 and 0.80? *Of course,* you may think -the former interval is substantially narrower than the latter, so our estimate of the population parameter is more precise. That may be so, but can we also be equally *confident* that our estimate will in fact be somewhere in the first interval as we can that it will be in the second? In other words, is the risk of stating that the parameter is somewhere within the interval *when in fact it is not* the same for both cases? Probably not.

In attempting any kind of interval estimation, the first thing we have to decide is the **confidence level** for our estimate, i.e. we have to decide on how often we want to be correct that our interval will in fact contain the population parameter in question. Different samples are likely to yield different values for the sample proportion P (see discussion of sampling error earlier) and, therefore, will produce different interval estimates of the population proportion π. Some of these intervals will contain π but others will not (see Figure 9.1). As the theoretical sampling distribution for a proportion has known properties, it will enable us to set k (i.e. the number of standard errors) in such a way that any specified percentage of these intervals will contain the desired parameter π (see also Chapter 8). Therefore, if k is set so that, say, 95% of all possible intervals under repeated sampling would contain π, then we have a 0.95 probability of selecting one of the samples that will produce an interval containing π. This interval, computed from the specific sample at hand, is known as a **95% confidence interval** and its upper and lower limits are denoted as **95% confidence limits**.

confidence level

95% confidence interval
95% confidence limits

In simple terms, a confidence interval reflects a range of values that we are confident (but not certain) contains the population parameter. Clearly, the wider the confidence interval, the more confident we can be that the particular interval will contain our (unknown) parameter; conversely, the narrower the interval, the less confident we can be that it includes the parameter (but *if* it does, then our estimate will be more precise). Ideally, of course, we prefer a narrow interval with a high degree of confidence.

Figure 9.1 Confidence intervals for several samples around parameter value π.

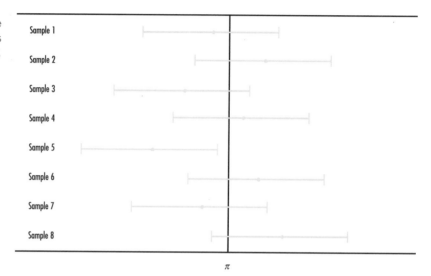

In general, a $100(1-\alpha)\%$ confidence interval for any population parameter Θ takes the following form:

$$\text{Sample statistic} - k_c(\text{standard error}) \leq \Theta \leq \text{sample statistic} + k_c(\text{standard error})$$

critical value
confidence coefficient

where k_c is known as the **critical value** (or 'reliability coefficient') corresponding to the **confidence coefficient** of $(1-\alpha)$. This critical value indicates the number of standard errors needed to create the desired confidence interval around Θ. Obviously, as α increases, the confidence coefficient decreases (e.g. if α goes from 0.05 to 0.10, the confidence coefficient reduces from 0.95 to 0.90) and, therefore, k_c also decreases; this results in a narrower interval for Θ (but accompanied by a reduction in confidence). Conventional values for α are 0.01, 0.05 and 0.10, resulting in 99%, 95% and 90% confidence intervals; however, there is nothing sacred about these values and if you want to set up an 83% or 74% confidence interval, nobody can stop you.

People often get confused when trying to interpret a confidence interval. A typical mistake is to make statements such as 'the probability is 0.95 that the parameter in question is included in the interval'. This is clearly not correct because the probability that the parameter is within a *particular* interval is either 1 or 0; the parameter in question is either included in the interval or it is not (see Figure 9.1). The correct interpretation of a confidence interval emphasizes the *procedure* used to generate it, not the specific interval obtained. More specifically, we can say that the procedure is such that in the long run 95 per cent of the intervals obtained will include the true (fixed) parameter. You should be careful not to assume that the particular interval you have obtained has any special property not possessed from comparable intervals that would be obtained from other samples.[2]

Having said all this, making a statement such as 'we are 95% confident that the particular interval we have constructed contains the population parameter' is actually OK, as long as we are clear that what we *really* mean is that '95% of all possible intervals constructed in this way will include the parameter concerned' (and, thus, our particular interval has a 0.95 chance to be one of them).

So how about putting all this seriously theoretical stuff into practice? By now, you must be *dying* to estimate the proportion of doctors who hate going to the dentist (and throw in a confidence interval for good measure). Let's do it.

HINT 9.1

When interpreting a confidence interval, always remember that the particular interval is only *one* out of many possible ones based on different samples; hopefully, it is one of those intervals that *does* include the population parameter!

Estimating the Population Proportion

You will be pleased to know that the steps we shall follow below are not specific to the population proportion but reflect a general sequence that can be followed whenever you feel like estimating any population parameter. Thus, if you understand the basic logic and decisions involved in estimating the population proportion, you will be in a position to apply your knowledge to estimate means, variances, and the like (and thus discover a whole new way of spending your weekends!).

To refresh your memory, the specific estimation problem we face is that of estimating the proportion of medical doctors who hate going to the dentist. Thus, our unknown population parameter is π and we would like to develop, say, a 95% confidence interval for it. We further know from our (random) sample of 450 doctors that 63% of them hate visits to the dentist, so $n = 450$ and $P = 0.63$.

What we must do now is find the appropriate sampling distribution for our sample statistic. This can be quickly achieved by consulting any standard statistics textbook (see Further Reading section for some readable examples) or by accosting your local friendly statistician. It just so happens that, with large samples, the sampling distribution of the proportion approximates a normal distribution (hurrah, we know this one!), with mean equal to the population proportion π and standard deviation equal

to $\sqrt{\dfrac{P(1-P)}{n}}$. The latter is, therefore, our beloved standard error s_p (see earlier), and comes to

$$\sqrt{[0.63(1-0.63)/450]} = \sqrt{[(0.63)(0.37)/450]} = 0.023$$

All we need to do now is set our critical value k_c and then we have all the information required to construct our confidence interval shown below (note that this is merely the application of the general formula for confidence intervals with respect to the population proportion):

$$P-k_c s_p \leq \pi \leq P+k_c s_p$$

Since we want a 95% confidence interval and we know that the sampling distribution involved approximates a normal distribution, our problem is to determine a value for k_c such that 95% of the area under the normal curve will be taken up by the interval. Yes, we *have* done this before in Chapter 8. And yes, you are right that we can get the value for k_c simply by going to the standard normal distribution tables which tell us, at a glance, how many standard deviations either side of the mean enclose 95% of the total area. And yes, you are right again if you think that k_c can be directly determined by looking at the z-values (i.e. standard scores) defining this area.

Indeed, if we rush to our statistical tables we should find that 95% of the area under the standard normal distribution lies between $z = -1.96$ and $z = 1.96$ (see Figure 9.2(a)). Thus if we set $k_c = 1.96$, we've cracked it.

Figure 9.2   95% and 99% confidence intervals: normal distribution.

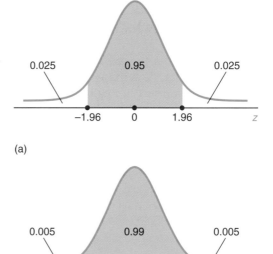

(a)

(b)

Applying the formula for a confidence interval to our data gives us the following result:

$$0.63-1.96(0.023) \leq \pi \leq 0.63+1.96(0.023) \quad \text{or} \quad 0.584 \leq \pi \leq 0.675$$

Thus we can conclude with 95% confidence that the proportion of doctors who hate going to the dentist is somewhere between 58.4% and 67.5%. End of story.

So, what if we want to set a 99% confidence interval instead? Do we have to go through the entire set of calculations and bore ourselves to death again? Not at all. Everything stays the same, the only bit of information that needs changing being k_c (in order to correspond to our new confidence coefficient). Back to our trusted standard normal distribution tables, we look for z-scores on either side of the mean such as that 99% of the total area under the curve is covered. These come to $z = -2.58$ and $z = 2.58$, so setting $k_c = 2.58$ gives us the following 99% confidence interval (see also Figure 9.2(b)):

$$0.63 - 2.58(0.023) \leq \pi \leq 0.63 + 2.58(0.023) \quad \text{or} \quad 0.570 \leq \pi \leq 0.689$$

Thus we can be 99% confident that between 57% and 68.9% of doctors hate going to the dentist. Note that this interval is wider than the corresponding 95% interval, reflecting the trade-off between the increase in confidence and loss in precision discussed in the previous section. Needless to say, we could follow the above procedure to set intervals at *any* confidence level – see how handy the standard normal tables are?

If only to get statistical purists off our back, we need to quickly mention that estimating the population proportion using the normal distribution as we have done above is only legitimate when we have a reasonably sized sample and when P is neither too close to 0 nor to 1. A rule of thumb often used to check these requirements is that both nP and $n(1-P)$ are greater than 5 (in our case they come to 283.5 and 166.5, so no problems). If nP and/or $n(1-P)$ are less than 5, then the appropriate sampling distribution is not the normal but the **binomial distribution**, which is the probability distribution for a dichotomous variable (we will come across the binomial distribution again in Chapter 11). However, as the sample size increases, the binomial distribution approximates the normal (a phenomenon that statisticians proudly describe as the 'normal approximation to the binomial'); it is for this reason that, with large samples, it is OK to go ahead and use the normal distribution as the appropriate theoretical sampling distribution for estimating a proportion.

binomial distribution

We should also mention that if the sample constitutes more than 5% of the population, we should use what is called a **finite population correction** in our calculation of standard error. This correction is defined as

finite population correction

$$\sqrt{\frac{N-n}{N-1}},$$

where N = population size and n = sample size. Thus the standard error of the population proportion corrected by this factor is

$$s_p = \sqrt{\left(\frac{P(1-P)}{n}\right)} \times \sqrt{\left(\frac{N-n}{N-1}\right)}$$

Note that, in most practical applications, the samples taken are usually a small fraction of the entire population and, thus, the finite population correction can safely be ignored.

Estimating the Population Mean

The procedure one uses to estimate a population mean μ is identical to the one followed above to estimate the population proportion. The only difference comes in the calculation of standard error, as means and proportions are different statistics and, not surprisingly, require different formulae.

sampling distribution of
the mean

If the population is normally distributed (and there are tests to establish this, as we will see in Chapter 11), the **sampling distribution of the mean** is also normally distributed with mean equal to the population mean μ and standard deviation equal to $\sigma/\sqrt{n}$, where σ is the *population* standard deviation and n is the sample size. Given that in most instances we do not know the standard deviation of the population, this version of the sampling distribution of the mean is as useful for practical estimation problems as a copy of *Gardeners' Weekly*! Luckily, if the sample is large enough (most experts suggest $n>30$), the sampling distribution of the mean can be approximated by a normal distribution with mean μ and standard deviation $s/\sqrt{n}$, where s is the *sample* standard deviation. What is even better is that this version of the sampling distribution of the mean is applicable regardless of the form of the distribution in the population.

standard error of the
mean

Thus, with a decent sample, the **standard error of the mean** is $s_{\bar{x}} = s/\sqrt{n}$ and the sampling distribution is normal.

As an example, say that, in addition to estimating the proportion of doctors who hate going to the dentist, we also wanted to estimate the number of hours they spent at the golf course each week and set a 90% confidence interval. Say that from our sample of 450 doctors we found that the mean number of hours per week spent at the golf course was 9.5 with a standard deviation of 6.8. All we need to do is to fill in the numbers in the following (by now surely familiar) formula:

$$\bar{x}-k_c s_{\bar{x}} \leq \mu \leq \bar{x}+k_c s_{\bar{x}}$$

Note that we have simply substituted the symbols for the sample mean, population mean and standard error of the mean in the general formula for a confidence interval that we introduced earlier in this chapter. There is nothing new here. Given that our sample size is large (well over 30), we can use the normal distribution as the appropriate theoretical sampling distribution for the mean and calculate the standard error as

$$s_{\bar{x}} = s/\sqrt{n}$$

This comes to $6.8/\sqrt{450} = 0.32$. Given that our confidence coefficient is 90% and the sampling distribution involved is normal, we can go to the standard normal tables (exactly as we did before when we estimated a proportion) and find the z-scores between which 90% of the total area under the curve is enclosed; these come to $z = -1.64$ and $z = 1.64$ (see Figure 9.3). We now set $k_c = 1.64$ and we have our desired confidence interval:

$$9.5-1.64(0.32) \leq \mu \leq 9.5+1.64(0.32) \quad \text{or} \quad 8.98 \leq \mu \leq 10.02$$

Thus we can be 90% confident that the mean number of hours (per week) spent at the golf course by doctors is between 8.98 and 10.2 hours (shame on them, they should be out there curing people!).

What happens if we do not know the population standard deviation *and* our sample is small? Well, not all is lost. If the distribution in the population is approximately normal then, instead of the normal distribution, we

t-distribution

should use the **t-distribution** (with $n-1$ degrees of freedom) to set k_c (the formula for calculating the standard error $s_{\bar{x}}$ stays the same).

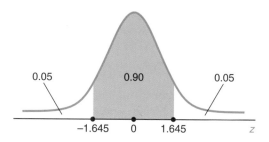

Figure 9.3 90% confidence interval: normal distribution.

The *t*-distribution is a symmetrical distribution like the normal distribution and almost as beautiful but a bit more platykurtic, i.e. less peaked at the centre and higher in the tails; as the sample size increases, the *t*-distribution approaches the normal (at about $n = 120$, the two distributions are practically identical). The *t*-distribution is really a 'family of distributions', since there is a different *t*-distribution for each different number of degrees of freedom; the latter are given by $n-1$, where n is the sample size. There are statistical tables available for the *t*-distribution (for different degrees of freedom) showing the area under the curve corresponding to different **t-values** (or '*t*-statistics'); these are equivalent to *z*-scores in the standard normal distribution and their interpretation is very similar. However, published tables for the *t*-distribution are nowhere as detailed as those for the standard normal distribution, as *t*-values depend on the degrees of freedom; typically, only the *t*-values enclosing 90%, 95% and 99% of the total area under the curve are reported for a range of degrees of freedom. Note also that, since the *t*-distribution approaches the normal as the sample size increases, for practical purposes, the *t*-distribution is often substituted by the standard normal distribution when $n>30$ (with 30 degrees of freedom the *t*-values enclosing 95% of the area come to -2.04 and 2.04, while with 120 degrees of freedom, the corresponding values are -1.98 and 1.98; given that the equivalent *z*-scores are -1.96 and 1.96 (see Figure 9.2(a)), the loss in precision is trivial if the standard normal distribution is used instead of the *t*-distribution once the sample size exceeds 30).

t-values

Here is a quick example of estimating the population mean using the *t*-distribution with a small sample. Suppose that we have asked 20 randomly selected primary school children to tell us the number of fancy-dress parties they have attended during the past month and we want to estimate the mean 'fancy-dress party attendance' with 95% confidence (clearly a profound social issue!). Assume that we have calculated the mean and standard deviation from our sample data and we got 14.6 and 3.9, respectively. Now, we cannot use the normal distribution as our sampling distribution because not only do we not know the standard deviation of the population (see earlier) but our sample is small as well. However, assuming that the population is normally distributed (or roughly so), we can use the *t*-distribution (with $20 - 1 = 19$ degrees of freedom) to find the critical value k_c. As before, we calculate the standard error of the mean as

$$s_{\bar{x}} = s/\sqrt{n}$$

This gives us $3.9/\sqrt{20} = 0.87$. Next, we go to the tables of the *t*-distribution (for 19 degrees of freedom) and look up the *t*-values such that 95% of the area under the curve is enclosed between them; these come to $t = -2.093$ and $t = 2.093$ (see Figure 9.4). So we set $k_c = 2.093$ and we get

$$14.6-2.093(0.87) \leq \mu \leq 14.6+2.093(0.87) \quad \text{or} \quad 12.78 \leq \mu \leq 16.42$$

Therefore, we can be 95% confident that, on average, primary school children attended between 13 and 16 fancy-dress parties over the past month.

Figure 9.4 95% confidence interval: *t*-distribution (df = 19).

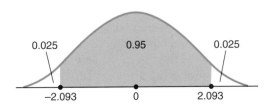

Now, if only to play devil's advocate, you might ask: what is the sampling distribution of the mean if (a) we do not know the standard deviation of the population, (b) the sample size is small, *and* (c) the population distribution is not normal? The 'formal' answer to this question is that, under these conditions, we cannot specify a probability distribution for the fluctuation of the mean from sample to sample to enable accurate estimation to take place (i.e. we are really and truly stuck!). The 'informal' answer is that you should have got a decent-sized sample in the first place and then you would not be asking such awkward questions!

Note that the comments made in the previous section concerning the application of the finite population correction $\sqrt{(N-n)/(N-1)}$ when the sample is more than 5% of the population also apply here. We must adjust the standard error of the mean in a similar fashion as we did for the standard error of the proportion; the correction factor formula stays the same.

Estimating Other Population Parameters

While the estimation of the population mean and the population proportion are the most frequent estimation applications in practice, one can follow similar procedures to estimate other parameters, such as the variance of a population, the difference between means in two populations, the difference between two population proportions, or the ratio of two variances. We see little point in going through what are essentially identical steps to demonstrate the setting of confidence intervals for such parameters (it would be incredibly boring and add little to what you already know). Instead, we have compiled in Table 9.1 a neat little summary of the general steps one has to go through in order to estimate any parameter, and we refer you to the Further Reading section for more specific details (i.e. appropriate sampling distributions, computation formulae for standard errors, and relevant statistical tables).

Computer packages can be of considerable assistance when you go through the steps outlined in Table 9.1. Some will provide you with the standard error of a requested statistic as a matter of course, while others

1. Decide on which population parameter, Θ, you want to estimate; this could be a proportion, a mean, a variance, etc.
2. Decide on the confidence coefficient $(1-\alpha)$ for your estimate of Θ and set your confidence level $100(1-\alpha)\%$.
3. Calculate the corresponding statistic (i.e. sample mean, proportion, etc.) from your sample data; needless to say this has to be a *random* sample, otherwise estimation goes out of the window as sampling error cannot be assessed (see Chapters 2 and 6).
4. Find the appropriate sampling distribution of the statistic and calculate its standard deviation; this is your standard error.
5. By looking at the tables of the sampling distribution, determine the critical value, k_c, corresponding to your confidence coefficient $(1-\alpha)$.
6. Construct your $100(1-\alpha)\%$ confidence interval by multiplying the critical value, k_c, with the standard error and subtract/add this product to the sample statistic.
7. Go to the pub for a well-deserved drink.

Table 9.1   Steps in estimation

may even set confidence intervals for you (at least for some parameters). While such facilities mean that you have to spend less time constructing confidence intervals, they do not reduce the burden of interpreting confidence intervals – *you* still have to do this, and frequent reference to Hint 9.1 should ensure that you do it correctly.

Before we leave this chapter, we should draw your attention to an interesting feature of the formulae for standard errors, namely that the sample size is always included in the denominator. Thus, the larger the sample size, the smaller the standard error. Given that the standard error is always multiplied by the critical value, k_c, when setting a given confidence interval (see earlier sections), the smaller the standard error, the narrower will be the interval (all other things being equal, of course). Thus, at a given level of confidence, increasing the sample size results in a narrower (i.e. more precise) interval estimate of the population parameter. This is worth bearing in mind when determining the sample size of your study.

Finally, we should mention that all the procedures and formulae presented in this chapter assume that a **simple random sample** has been drawn (see Table 2.2 in Chapter 2 for a description of different sampling methods). The calculation of standard errors and confidence intervals under alternative sampling methods (e.g. stratified or cluster sampling) is more complex and we refer you to the Further Reading section and the sources recommended in Chapter 2 for more details.

HINT 9.2

While (sample) size is not everything, in estimation, it helps!

simple random sample

Summary

In this thoroughly titillating chapter, we looked at how we can say something about the population when we only have a sample. We began by distinguishing between point estimates and interval estimates, and highlighted the importance of understanding the fluctuation of sampling error. Next, we introduced the concept of a sampling distribution and indicated how its standard deviation can be used to represent the variation in sampling error for the statistic concerned. We then considered issues associated with the construction and interpretation of confidence intervals and provided illustrations involving different population parameters and different sampling distributions. Estimation was our first close encounter with the inferential branch of statistical analysis; we will be using our knowledge in the next chapter, in which statistical inference will be approached from a different angle, namely that of hypothesis-testing.

Questions and Problems

1. What is the general objective of statistical estimation?

2. What is the difference between a population parameter and a sample statistic?

3. What is the difference between a point estimate and an interval estimate?

4. What is a sampling distribution? Why is it useful?

5. Explain the difference between standard error, standard deviation of the sample and standard deviation of the population.

6. What does the term 'confidence limits' mean? How would you go about setting a confidence interval for the population mean?

7. Explain what the 'critical value' (reliability coefficient) indicates. Give an example to illustrate your answer.

8. Under what circumstances would you choose the *t*-distribution in preference to the standard normal distribution?

9. What role does the size of the sample play in estimation?

10. What would be your best estimate of the average mental age of the authors?

Notes

1. Blalock, HM (1979) *Social Statistics*, revised 2nd edn, p. 106. Tokyo: McGraw-Hill Kogakusha.
2. Blalock, HM (1979) *Social Statistics*, revised 2nd edn, p. 211. Tokyo: McGraw-Hill Kogakusha.

Further Reading

Churchill, GA (1995) *Marketing Research: Methodological Foundations*, 6th edn. Fort Worth, TX: Dryden Press. Chapters 10 and 11 contain good discussions of estimation issues and highlight the role that sample size plays when setting confidence intervals.

Daniel, WW & Terrell, JC (1995) *Business Statistics for Management and Economics*, 7th edn. Boston: Houghton Mifflin. You will probably faint when you see the size of this book (the Appendices alone take up more than 200 pages!), but it is surprisingly easy to follow. More of a handbook than a pocket companion, it covers not only the methods covered in this chapter but also most of the tests and techniques described in Chapters 10 to 14. It also has plenty of computer-based examples using the *MINITAB* and *STAT+* statistical analysis packages.

Tull, DS & Hawkins, DI (1993) *Marketing Research: Measurement and Method*, 6th edn. New York: Macmillan. If you want to know quickly and painlessly how big your sample must be to estimate a population mean and/or proportion given a level of confidence *and* a specification of maximum error (i.e. width of interval), the 'nomographs' in Chapter 16 are invaluable.

How about sitting back and *hypothesizing?*

The Nature and Role of Hypotheses

In the last chapter we introduced basic techniques for estimating population parameters from sample statistics. A complementary approach to making inferences about the population is via **hypothesis-testing**. Whereas in estimation the focus was on making some 'informed guesses' about the values of population parameters using our sample data and a relevant sampling distribution, in testing hypotheses the aim is to examine whether a particular proposition concerning the population is likely to hold or not. For example, an advertising executive may hypothesize that placing perfume ads in *Heavy Metal Weekly* will attract more readers than placing them in *Shipbuilders' Digest*. A personnel manager may hypothesize that employees who have undergone a 'find your elusive inner-self' training programme (offered by *Weird Solutions Inc.*) will be more productive than employees who have not. Or an economist may hypothesize that the average personal income not declared to the Inland Revenue by tax collectors is £1,200 per annum. In short, a hypothesis is a statement regarding a population (or populations) that may or may not be true.

Let us assume that, on the basis of past evidence, theory, or seven years of conceptual deliberation, we hypothesize that the Dutch are more likely to own bicycles than the Vietnamese; this is simply a *conjecture* on our part which may or may not be true in the population. Now, if we had unlimited resources in terms of time and money we could undertake a census in the Netherlands and a census in Vietnam and determine whether our hypothesis is correct or not. However, as most of us do not have unlimited resources (otherwise we would be spending our time and incredible wealth on better things), what we have to do is make inferences as to the correctness of our hypothesis from *sample* information. If we took random samples of Dutch and Vietnamese consumers and found, say, that 57% of Dutch consumers owned bicycles as compared to 28% for the Vietnamese, our conjecture would find support in the data. If, on the other hand, we found that the proportions were 15% and 36%, respectively, our conjecture would be refuted by the data (indeed, it would appear that the opposite would hold). Finally, if we found that, in both samples, exactly 34% owned bicycles, again our conjecture would not find support (as both country samples show identical ownership rates).

This is all very nice, but what if the figures do not turn out as clear-cut as that? Can we really say that we found support for our hypothesis if the percentages turn out to be, say, 29% for the Dutch and 28% for the Vietnamese? And, equally, should we immediately conclude that the 'true' situation in the population is exactly the opposite from what we predicted if the proportion for the Dutch sample is 36.4% and that for the Vietnamese sample 36.5%? If you feel uncomfortable about drawing such conclusions, well you should! The problem, as you should know by now, is the dreaded sampling error. As was the case with estimation, data from a sample are just that: data from a sample. Any quantity (such as the per-centages calculated here) based on sample data is subject to sampling error which is not constant but variable (see Chapter 9). Unless we somehow incorporate sampling error in our deliberations, it becomes very difficult to determine whether our hypotheses are supported by the data or not. Thus, conceptually, the problem we are facing is very similar to that faced when estimating any population parameter: we must consider sampling error whenever we make inferences about the population. Before we show you how this is done, it is important to look at the notion of a hypothesis from several different angles.

To begin with, every time we set a hypothesis we are really setting *three* of them. No, we have not been drinking – any hypothesis implies two other hypotheses, as we shall immediately demonstrate beyond reasonable doubt. Take our previous hypothesis (let's call it H_1), which states:

H_1: *A greater proportion of Dutch consumers own bicycles than of Vietnamese consumers*

This hypothesis indicates what we think the situation is in the population. By implication, we do *not* think that

H_2: *A smaller proportion of Dutch consumers own bicycles than of Vietnamese consumers*

or that

H_3: *The proportions of Dutch and Vietnamese consumers who own bicycles are the same*

Any *one* of H_1, H_2, or H_3 could reflect the actual situation in the popu-lation. Obviously, we do not know which one (otherwise why bother with hypothesis-testing?), but we think that H_1 is correct. Note that nothing would change if our initial hypothesis was H_2 (or H_3 for that matter); the other two hypotheses would be 'automatically' generated. Why this happens is quite simple: the three hypotheses constitute *mutually exclusive* and *collectively exhaustive* descriptions of all possible situations in the popu-lation. *Either* the Dutch are more likely to own bicycles, *or* the Vietnamese are more likely to own bicycles, *or* the Dutch and the Vietnamese are equally likely to own bicycles; there are no other possibilities (if you can come up with additional possibilities then you deserve a Nobel prize and we deserve to be shot!).

competing hypotheses

From the above it should be clear that whenever we examine a hypothesis, we are really comparing our hypothesis against two other **competing hypotheses**; consequently, support for our hypothesis implies rejection of the other two hypotheses. However, in practice, what we usually do is test one hypothesis against a combination of the other two (it simply makes life easier). Thus, if our interest is in H_1, we can combine H_2 and H_3 into a new hypothesis (call it H_4) as follows:

> H_4: *The proportion of Dutch consumers who own bicycles is smaller than or equal to that of Vietnamese consumers*

What if our intial hypothesis is H_2 instead of H_1? No problem. We formulate another hypothesis (say H_5) by combining H_1 and H_3:

> H_5: *The proportion of Dutch consumers who own bicycles is larger than or equal to that of Vietnamese consumers*

null hypotheses

Note that both H_4 and H_5 include the possibility of no difference between the two groups. Hypotheses of this kind are known as **null hypotheses**; more specifically,

> the term null hypothesis reflects the concept that this is a hypothesis of no difference. For this reason, the null hypothesis always includes a statement of equality. When it is presented symbolically, it contains an equals sign.[1]

If you go back to H_3 you will see that it too is a null hypothesis; the competing hypothesis it implies is simply the combination of H_1 and H_2 (call it H_6):

> H_6: *The proportions of Dutch and Vietnamese consumers who own bicycles are not equal*

alternative hypothesis

If you look again at hypotheses H_1 through to H_6, you will see that the null hypotheses are accompanied by other hypotheses which are not of the null-type; we call this hypothesis the **alternative hypothesis** (sometimes also known as the 'research hypothesis'). An alternative hypothesis is the complement of the null hypothesis, that is it postulates some difference or inequality; as such, it can *never* include a statement of equality. Professional statisticians tend to have fits if you confuse the null and alternative hypotheses, so please keep Hint 10.1 in mind.

Hypotheses H_3, H_4 and H_5 are all null hypotheses, while hypotheses H_6, H_1 and H_2 are their respective alternative hypotheses. Altogether, as Table 10.1 shows, we can form three pairs of hypotheses concerning the proportions of Dutch (π_1) and Vietnamese (π_2) consumers who own bicycles; note also that these three pairs cover all possible hypotheses that can be formulated concerning the population proportions π_1 and π_2.

Why should we bother distinguishing between null and alternative hypotheses? Does it really matter that some hypotheses are null hypotheses and others alternative hypotheses? The answer is an unqualified yes: it *does* matter and it matters a *lot*. The reason it matters is that we can never

HINT 10.1

Look for any mention of equality when reading a hypothesis; if you find one, then you are dealing with a null hypothesis; if not, then you are facing an alternative hypothesis.

	Case A	Case B	Case C
Null hypothesis	$\pi_1 \leq \pi_2$	$\pi_1 \geq \pi_2$	$\pi_1 = \pi_2$
Alternative hypothesis	$\pi_1 > \pi_2$	$\pi_1 < \pi_2$	$\pi_1 \neq \pi_2$

Table 10.1 Examples of null and alternative hypotheses

directly test an alternative hypothesis – as we shall see shortly, our hypothesis-testing procedures can only deal with null hypotheses; it is only when a null hypothesis is *rejected* as being untenable that we obtain indirect support for the corresponding alternative hypothesis. As we cannot test H_1, H_2 and H_6 directly, what we do is test their respective *null* hypotheses (i.e. H_4, H_5 and H_3); this is why we always form pairs of hypotheses (such as the ones shown in Table 10.1). Indeed,

> we never set out to prove anything; our aim is to show that an idea is untenable as it leads to an unsatisfactorily small probability. In other words tests are designed to disprove hypotheses . . . the hypothesis we are trying to disprove is always chosen to be the one in which there is no change . . . This is why it is usually referred to as the null hypothesis.[2]

There are good reasons why we have to follow this 'roundabout' procedure in hypothesis-testing; while a formal explanation is beyond the scope of this text, by the time you finish this chapter you should get a rough idea (see also Further Reading section).

Bearing the above in mind, if you compare the three alternative hypotheses in Table 10.1, you will notice that all three postulate the *existence* of differences in the proportions of Dutch and Vietnamese consumers who own bicycles. However, in cases A and B the *direction* of the difference is also specified, while in Case C it is not. Alternative hypotheses which, in addition to the existence of differences, also indicate the direction of the expected differences are known as **directional hypotheses**. In contrast, hypotheses which only postulate a difference without any *a priori* expectations as to the direction of the differences are called **exploratory hypotheses**.

The formulation of directional hypotheses presupposes greater prior knowledge about the issue at hand; such knowledge may be available from past theoretical work and/or empirical evidence. Exploratory hypotheses, on the other hand, are more 'cautious' or 'conservative' and reflect either a lack of prior knowledge or conflicting past findings; thus one is uncertain about the direction in which a difference will be manifested. For example, if you know that there had been four previous studies in the past 20 years and all had concluded that Dutch consumers are more likely to own bicycles than Vietnamese consumers, then it would not be inappropriate to formulate a directional hypothesis along the lines of Case A in Table 10.1. On the other hand, if two of the previous studies had shown a greater likelihood of ownership among the Dutch, one study had found no difference and yet another indicated that the Vietnamese are more likely to own bicycles than the Dutch, then an exploratory hypothesis (i.e. Case C) would be called for. An exploratory hypothesis would also be appropriate if no previous studies had been undertaken to enable you to formulate some directional expectations concerning bicycle ownership in the two countries

WARNING 10.1

Only null hypotheses can be tested – *if* they are rejected, this is taken to signify support for the alternative hypothesis. We can never test an alternative hypothesis directly and nor can we ever *prove* a hypothesis.

directional hypotheses

exploratory hypotheses

and you had little theory to rely on. Note that the distinction between directional and exploratory hypotheses has important implications for the actual testing of such hypotheses, as we will see in the next section.

So far we have been discussing null and alternative hypotheses relating to how two groups (in this instance, Dutch and Vietnamese consumers) may differ in terms of some characteristic (in this case bicycle ownership). Hypotheses concerning differences between two (or more) groups are only one of many types of hypotheses. Another type is hypotheses concerning differences between measurements. Such hypotheses are typically formulated in the context of an experimental design, in which the same group is compared on the same variable before and after administration of the experimental treatment. For example, one may hypothesize that rap music increases typing speed; to test this hypothesis one may compare the typing speed of a group of secretaries under a 'no music' (i.e. silence) condition versus a 'rap music' condition. As another example, not involving experimentation, consider a hypothesis suggesting that the leader of the *Exceedingly Conservative* party has a more positive image among voters in Great Yarmouth than the leader of the *Excessively Liberal* party. This could be tested by taking a random sample of voters and asking them to state their opinions for *both* leaders – again, the comparison would involve the same group but different measurements. Hypotheses involving comparisons between groups and differences between measurements will be examined in detail in Chapter 12.

A rather different type of hypothesis involves the examination of *relationships* (or 'associations') between variables. For example, a sociology professor may hypothesize that the higher one's income, the greater one's consumption of strawberry ice cream; here, a positive relationship between income and consumption is postulated, indicating a directional hypothesis. A directional hypothesis would also be indicated if a negative link had been specified between income and strawberry ice cream consumption. Had the direction of the relationship not been specified (i.e. income and strawberry ice cream consumption are related but we do not know *how*), then the hypothesis would have been an exploratory one.

Note that, strictly speaking, hypotheses concerning differences between groups could also be interpreted as implying a relationship; for example, postulating differences between Dutch and Vietnamese consumers in terms of bicycle ownership can be interpreted as hypothesizing a relationship between nationality and bicycle ownership. Be that as it may, we usually talk about relationships when 'positive' and 'negative' connotations are meaningful and readily interpretable. Thus, while a positive relationship between income and ice cream consumption states unambiguously that the more money one has the more ice cream he/she is likely to eat, a positive relationship between nationality and bicycle ownership is not as clear. The reason for this is that nationality is a *nominal* variable and, as such, it contains no meaning of order (see Chapter 3). Therefore, the scoring of this variable is arbitrary and, as a result, any 'relationship' between it and another variable is dependent upon the specific scoring used. Thus, a 'positive' link between nationality and bicycle ownership with Dutch = 1 and Vietnamese = 0, automatically becomes a 'negative' link if we score Dutch = 0 and Vietnamese = 1! However, the substantive nature of the link is clearly the same in both instances: Dutch consumers are more likely to own bicycles than Vietnamese consumers. In general, when a nominal variable is involved it is much better to state hypotheses in terms of differences and reserve the term 'relationship' for hypotheses involving variables measured at (at least) ordinal level. Hypotheses involving relationships will be examined at length in Chapter 13.

In addition to hypotheses regarding differences (between groups or measurements) and hypotheses concerning relationships, there are *single-variable* hypotheses. Such hypotheses can relate to postulated values of particular population parameters (e.g. that a mean, median or standard deviation is equal to a certain value), the shape of the distribution (e.g. that a variable follows a normal distribution), and the nature of the observations (e.g. that a set of observations is indeed random). Hypotheses of this kind will be the subject of Chapter 11. For your convenience, Figure 10.1 provides a 'route map' for the various types of hypotheses covered in the rest of this book.

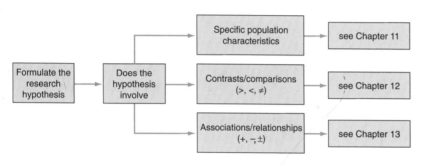

Figure 10.1 Types of hypotheses.

We are now ready to tackle the more 'technical' part of hypothesis-testing. In what follows, we shall go through the hypothesis testing procedure step by step, highlighting the main issues along the way. The principles of hypothesis-testing are the same irrespective of (a) the type of hypothesis under consideration, and (b) the specific statistical test involved. As long as you grasp the key concepts discussed in the next section, you should be able to deal with any hypothesis and have no problem whatsoever understanding the specific techniques presented in Chapters 11–13. Conversely, failure to devote sufficient time to understanding the basic principles underlying the testing of hypotheses means that you are likely to end up in tears, not least because 'it is very easy to become so overwhelmed with the details of each particular test encountered that one is unable to see the similarities underlying all tests'.[3]

A General Approach to Hypothesis-Testing

As Table 10.2 shows, there are five distinct steps associated with formulating and testing a hypothesis. While some of these steps will normally be carried out by your computer, it is imperative that you understand what is happening at each step and why. Consequently, we will take a hands-on approach here and illustrate the general principles involved by the use of a specific example without the use of a computer (shock, horror!); yes, this means that there will be some calculation involved, but it's for your own good!

Table 10.2 Steps in hypothesis-testing

1. Formulate the null and alternative hypotheses.
2. Specify the significance level.
3. Select an appropriate statistical test.
4. Identify the probability distribution of the test statistic and define the region of rejection.
5. Compute the value of the test statistic from the data and decide whether to reject or not reject the null hypothesis.

Step 1: Formulation of Null and Alternative Hypotheses

The first step in Table 10.2 is rather obvious; however, it is also extremely important because it largely determines what happens next. As already mentioned, the null hypothesis should contain a statement of equality (i.e. either $=$, $\geq$, or $\leq$). By convention, a null hypothesis is denoted as H_0 (read: H-nought) and is always given the benefit of the doubt, that is it is *assumed* to be true unless it is rejected as a result of the testing procedure. Note that inability to reject the null hypothesis does not *prove* that H_0 is actually true; it may be true, but our tests are only capable of disproving (but not confirming) a hypothesis. In this context, 'evidence compatible with a hypothesis is never conclusive, whereas contradictory evidence is sufficient to cast doubt on a hypothesis'.[4]

To illustrate this logic, consider a claim by a friend of yours that drinking a cup of coffee with three spoonfuls of pepper in it will cure any hangover within five minutes. Next time you have a bit to drink, you decide to test his hypothesis and drink this revolting concoction; within five minutes you feel fine! Now, would you, on the basis of a single trial, conclude with certainty that it was your friend's amazing remedy that cured your hangover? Probably not. Nevertheless, you may be encouraged enough to repeat 'the treatment' when you get your next hangover. Imagine now that after successfully treating 128 major hangovers within the prescribed time, this time your hangover does not go away within five minutes (but takes three agonizing days to clear!). This one incident is enough to invalidate the hypothesis that the remedy is effective within five minutes; it took one piece of contradictory evidence to reject this hypothesis while 128 instances of results in favour of the hypothesis were not sufficient to prove it.

If, as a result of testing, the null hypothesis is rejected, this is interpreted as signifying support for the alternative hypothesis which, again by convention, we denote as H_1 (read: H-one). Remember that since H_0 and H_1 are complementary, the alternative hypothesis should always include a statement of inequality (i.e. $\neq$, $>$ or $<$). Depending upon whether the alternative hypothesis is exploratory or directional in nature, the form of the null hypothesis will also vary (see Table 10.1 earlier); as we will see, this has important implications for Steps 4 and 5.

Okay, we've conceptualized enough. Let us illustrate some of the points discussed above by means of an example. Suppose that we wish to test a hypothesis concerning the average number of monthly visits to the hairdresser by teenagers. Having researched the voluminous literature on teenagers' habits, we have unearthed 148 empirical studies in 73 different countries indicating that, on average, teenagers go to the hairdresser between 2 and 13 times each month. None of the studies, however, has been conducted in the United Kingdom and we see this as a unique opportunity to make our mark in the scientific community by providing much-needed UK-based evidence. Bearing in mind the findings of previous research, we speculate that UK teenagers will visit the hairdresser *more than* three times a month. This is our alternative hypothesis (H_1), which means that our null hypothesis (H_0) is that the average number of visits is three times or less (in practical applications, you may generally find it easier to state the alternative hypothesis first – as was done here – and then 'derive' the relevant null hypothesis from it). Denoting the population mean with the beautiful Greek letter μ (as we did in Chapter 9), we can formally state our hypotheses as follows:

$$H_0: \mu \leq 3$$
$$H_1: \mu > 3$$

Now, what we want is to be able to reject H_0 in favour of H_1 (which will make us famous when we publish our findings in the prestigious *Journal of Hairdressing Research*). We thus set out to see whether we should or should not reject H_0 (always remembering that failure to reject does not mean that we have proved H_0 – see Warning 10.1 above). We first have to collect some data, so we take a random sample of 200 teenagers across the UK (recall from Chapters 2, 6 and 9 that we *must* have a probability sample in order to engage in statistical inference). Next, we interview our sample and we find that the mean number of visits to the hairdresser is 3.5 with a standard deviation of 2.4. Where do we go from here?

Step 2: Specification of Significance Level

Having formulated the null and alternative hypotheses, the next step is to specify the circumstances under which we will reject H_0 and not reject H_0. Bearing in mind that we do not know for sure whether H_0 is true or false in the population (otherwise there is no point in hypothesis-testing), a rejected H_0 may be true or false; similarly, a non-rejected H_0 may also be true or false. Thus there are four possible outcomes whenever we test any hypothesis, namely:

1. Not rejecting H_0 when H_0 is true (i.e. failing to reject a true null hypothesis).
2. Rejecting H_0 when H_0 is true (i.e. rejecting a true null hypothesis).
3. Not rejecting H_0 when H_0 is false (i.e. failing to reject a false null hypothesis).
4. Rejecting H_0 when H_0 is false (i.e. rejecting a false null hypothesis).

As Table 10.3 indicates, outcomes 1 and 4 are desirable, representing correct decisions on our part. On the other hand, outcomes 2 and 3 are undesirable, since they indicate that we got it wrong. Our decisions are erroneous since they do not correspond with the true situation in the population. However, the two types of error are very different. What we term **Type I error** occurs if we reject the null hypothesis when we should not (i.e. when the null hypothesis is, in fact, true). In contrast, **Type II error** occurs if we do not reject the null hypothesis when we should reject it (i.e. when the null hypothesis is false).

Type I error
Type II error

If only to improve our knowledge of the Greek alphabet, let us call the probability of committing a Type I error α and the probability of committing a Type II error β; as H_0 can be *either* true *or* false (but not both), the probabilities of making a correct decision are $1-\alpha$ and $1-\beta$, respectively. Our problem is that since we do not know in advance whether H_0 is true or false, we need to guard against the possibilities of making *both* a Type I *and* a Type II error. Thus, ideally, we would like to keep α and β as low as possible so as to maximize the probabilities of reaching a correct decision. Unfortunately, there is a snag: for a fixed sample size, decreasing α results in an increase in β and decreasing β results in an increased α. Only by increasing the sample size will both errors decrease (the exact relationship between α and β is quite complex and discussing it here will probably confuse rather than enlighten; for more details, see the Further Reading section).

Table 10.3 Errors in hypothesis-testing

Decision made	Situation in the population	
	H_0 is true	H_0 is false
H_0 is not rejected	Correct decision $(1-\alpha)$	Type II error (β)
H_0 is rejected	Type I error (α)	Corrrect decision $(1-\beta)$

Bearing the above in mind, we are faced with a dilemma: should we minimize the risk of committing a Type I error (i.e. keep α as small as possible) or should we be more concerned with making a Type II error (and thus opt for a low level of β?). The way that we think about resolving this dilemma is best illustrated by the famous 'judicial analogy' shown in Table 10.4. Here, the two possible true conditions are paired with the two possible verdicts. Ideally, we want to find the defendant innocent when he/she is in fact innocent and guilty when he/she is in fact guilty; thus the top left and bottom right cells of Table 10.4 indicate correct decisions. What about the implications of getting it wrong? Are the two 'error cells' (i.e. bottom left and top right) equally undesirable? In other words, can we view the condemnation of the innocent in exactly the same way as the release of the guilty? Most people would not: finding an innocent person guilty when he/she in fact is innocent is seen as a much graver error than failing to convict a guilty person. Indeed, the judicial principle of 'innocent until proven guilty' that characterizes modern law is based upon the recognition that the two types of error in Table 10.4 are not of equal importance; making a Type I error is seen as a much more serious failure of the judicial process than making a Type II error.

Recalling that in hypothesis-testing the null hypothesis is always given the benefit of the doubt (see step 1 earlier), we have an 'innocent until

Verdict	True situation: Defendant is	
	Innocent	Guilty
Innocent	Correct decision $(1-\alpha)$	Wrong decision (β)
Guilty	Wrong decision (α)	Correct decision $(1-\beta)$

Table 10.4 Judicial analogy of Type I and Type II errors

proven guilty' situation. The implication is that we should be particularly careful not to reject H_0 unless we have strong evidence against it. This, in turn, suggests that we need to minimize the risk of wrongly rejecting H_0, i.e. the probability of committing a Type I error, α. We denote α as our **significance level** and use it to indicate the maximum risk we are willing to take in rejecting a true null hypothesis: the less risk we are willing to assume, the lower the α. Typical values for α are 0.10, 0.05, 0.01 and 0.001; however, these are largely arbitrary and reflect tradition as much as anything else.

significance level

As was the case with interpreting a confidence interval (see Chapter 9), confusion often reigns in the interpretation of a significance level. Among the most common misinterpretations is that the significance levels shows the probability that the results occurred by chance; that one minus the significance level shows the confidence we can have in the alternative hypothesis; and that the significance level shows the probability that the null hypothesis is true. None of these interpretations is even remotely right. You should always associate a significance level with *a probability of making a mistake* and, at that, a particular kind of mistake: rejecting the null hypothesis when you shouldn't reject it (because it is true). Thus when we select the 5% significance level (i.e. set $\alpha = 0.05$) to conduct a hypothesis test, what we are saying is that we will conduct our test in such a way that we will only reject the null hypothesis when in fact it is true 5 times out of 100. In our example of teenager visits to the hairdresser, if the true number of visits is 3 or less (as our null hypothesis specifies), we would wrongly conclude that the number of visits is *not* 3 or less only 5 times out of 100; in other words, we would make a wrong ruling against a true null hypothesis relatively rarely.

> **HINT 10.3**
>
> Always remember that the significance level is a probability of making a mistake: rejecting a true null hypothesis.

Having specified a significance level, the way we use it is simple. If the result of our statistical test is such that the value obtained has a probability of occurrence less than or equal to α, then we reject H_0 in favour of H_1 and we declare the test result as *significant*. If, on the other hand, the probability associated with the test result is greater than α, then we cannot reject H_0 and we denote the test result as *non-significant*. Thus, by using a particular significance level, we can 'partition' all possible test results into (a) those that lead us to reject the null hypothesis (and, thus, indirectly support the alternative hypothesis), and (b) those that prevent us from rejecting the null hypothesis.

Note that in hypothesis-testing we use a statistical test in conjunction with a significance level to decide on whether or not to reject a null hypothesis. This is the reason why such statistical tests are often referred to as **significance tests** and hypothesis-testing as **significance-testing**. Indeed,

significance tests
significance-testing

a test of significance assesses the strength of the evidence against the null hypothesis in terms of probability. If the observed outcome is unlikely under the supposition that the null hypothesis is true, but is more probable if the alternative hypothesis is true, the outcome is evidence against H_0 in favour of H_1. The less probable the outcome is, the stronger the evidence against H_0 . . . if we take $\alpha = 0.05$, we are requiring that the data give evidence against H_0 so strong that it would happen no more than 5% of the time (one time in twenty) when H_0 is really true. If we take $\alpha = 0.01$, we are insisting on stronger evidence against H_0, evidence so strong that it would appear only 1% of the time (one time in a hundred) if H_0 is really true.[5]

Let us go back to our example of teenage visits to the hairdresser and agree on $\alpha = 0.05$ as our significance level. Note that we should really decide on the size of α *before* we collect any data (otherwise, by 'tweaking' α after we know what the sample data looks like, we can influence whether H_0 is rejected or not – see also Step 4 below). Recall that the mean number of monthly visits to the hairdresser came to 3.5. Our problem is to determine the probability of obtaining such a result under the assumption that the null hypothesis is true (i.e. that the actual number of visits to the hairdresser is 3 or less). If this probability is lower than or equal to 0.05 (our significance level) we could reject the null hypothesis and rule in favour of our alternative hypothesis (i.e. that the number of visits is more than 3). If, on the other hand, the probability is greater than 0.05 we cannot reject the null hypothesis and, thus, we have no evidence in support of the alternative hypothesis. In either case, to determine this probability we need some sort of statistical test; the selection of an appropriate statistical test is the next and extremely important step in hypothesis-testing.

Step 3: Selection of an Appropriate Statistical Test

A statistical test is simply a technique which can be used to test a particular hypothesis. There is little doubt that the selection of an appropriate statistical test is the most difficult and frustrating step in hypothesis-testing. The reason is simple: there are so many of them. Choosing among the multitude of tests currently available for testing every conceivable type of hypothesis is not exactly an enjoyable activity and can put you off data analysis for life! Unfortunately, it is a necessary evil, because if you select the wrong test for your type of hypothesis and/or data, the rest of the hypothesis-testing procedure goes out of the window (i.e. your results, significant or not, will be meaningless).

Before you get totally depressed, let us cheer you up by mentioning that, in Chapters 11 to 13, we have already selected for you those tests that you are most likely to need. Moreover, not only do we show you *how* to use *which* test for *what* type of hypothesis but also highlight the key assumptions and limitations underlying individual tests so that you avoid the pitfalls involved (what more could you possibly ask for?). Consequently, in this section we discuss only the broad considerations involved in choosing an appropriate test for your needs and provide a specific illustration applied to our example of teenager visits to the hairdresser.

Several factors enter into the choice of a statistical test. Many of these have already been mentioned in Chapter 6 in the context of the broader discussion of the influences affecting the choice of the method of analysis (so it would not harm you to go back and have another look at Chapter 6 and, in particular, Figure 6.1). Nevertheless, it is useful to look at choice criteria as relating specifically to techniques appropriate for hypothesis-testing purposes.

First, the *type* of hypothesis to be tested will require a different test each time; for example, different tests are appropriate for hypotheses concerning differences between groups compared with those for hypotheses concerning relationships between variables (imagine the unemployment rate among statisticians if there were one test you could apply to all possible hypotheses!). Moreover, even within a broad class of hypotheses, different tests may be required; for example, a different test is appropriate for testing differences between two means than from, say, between two medians. Finally, even for exactly the same hypothesis, a different test may be needed depending upon the size of your sample; for example, testing a hypothesis about a proportion with a large sample is done with a different test from that used with a small sample.

Second, the *distributional assumptions* made regarding the population(s) from which the sample(s) was drawn will affect the choice of test. A typical assumption, in this context, is that the sample data have been drawn from a normally distributed population; this assumption permeates a great many statistical tests, including tests for hypotheses involving means, variances, proportions and correlations. Another assumption that often crops up in hypotheses involving comparisons is that the samples come from populations with equal variances; tests of mean differences across two or more groups usually have this requirement. As already mentioned in Chapter 6, statistical tests that make (often stringent) assumptions about the nature of the populations from which the sample data are drawn are known as **parametric tests**; tests which do not make such stringent assumptions are known as **non-parametric tests** or 'distribution-free' tests. Although minor violations of the distributional assumptions of many parametric tests do not totally invalidate their results if the sample is large enough, small samples coupled with substantial deviations from distributional assumptions and/or measurement requirements (see below) are a sure recipe for disaster.

parametric tests
non-parametric tests

Third, the *level of measurement* of the variable(s) involved in the hypothesis under consideration is also relevant. Some tests are appropriate for nominal data (e.g. when frequencies are compared across two groups), other tests are suitable for ordinal data (e.g. when two rank orders are related to one another), and still others should really only be used on interval/ratio data (e.g. when differences between means are examined). A point made in Chapter 6 and worth repeating here is that parametric tests should only be applied to metric data (i.e. at least interval) and that parametric tests work best with 'large' sample sizes (i.e. at least 30 observations per variable/group); a trip back to Figure 6.2 in Chapter 6 at this point is highly recommended.

Given the restrictive nature of parametric statistical tests, you may wonder why we bother with them. If non-parametric tests can do the job,

power of a test

where is the benefit of using their parametric brethren? The answer is provided by the fourth (and, you will be glad to know, last) consideration affecting the selection of a statistical test, namely its *power*. The **power of a test** is defined as the probability of rejecting the null hypothesis when it should be rejected (i.e. when it is in fact false). Reference to Table 10.3 shows that power is the complement of making a Type II error, that is it is equal to $1-\beta$. In our judicial analogy in Table 10.4, power is illustrated in the bottom right cell where the guilty defendant is indeed found guilty by the court. It is in terms of power that parametric tests have the upper hand over non-parametric tests. Specifically, given a sample size and a null hypothesis that could be tested by both a parametric and a non-parametric test, the former test will be more powerful in terms of its ability to detect and reject a wrong null hypothesis. This is not surprising, since the more restrictive the circumstances in which a test is expected to apply (i.e. the more assumptions that must be satisfied), the more sensitive the test will be if these circumstances apply (i.e. if the assumptions are met). Bearing in mind that the power of the test increases with increases in sample size (see earlier), a parametric test will be more *efficient* than a corresponding non-parametric test as a lower sample size will be required by the former to achieve the same power as the latter. What all this boils down to is that if the assumptions of a parametric test are satisfied (and some of them can actually be tested, as we will see in the next few chapters), opting for a non-parametric test makes little sense as one would be losing out on power. Moreover, many parametric tests are relatively **robust**, that is relatively insensitive to mild-to-moderate departures in their assumptions (particularly with a large sample), which further encourages their use. Note that the degree of robustness is *test-specific* – that is not all parametric tests are equally robust. Having said all that, if serious violations of the underlying assumptions are evident, turning a blind eye and persisting with the parametric test is a cardinal sin (punishable by having to memorize and then recite backwards the 138-volume *Encyclopaedia of Advanced Statistical Procedures!*).

robust test

WARNING 10.3

Do not let your lust for (statistical) power drive your test choice. A powerful test with violated assumptions is not powerful – it is simply invalid.

z-test

Let us go back to our example and try to apply some of the riveting issues discussed above. Here's what we have: we want to test a hypothesis about a mean, we have a nice large sample ($n = 200$), our data is ratio (number of monthly visits to the hairdresser) and, obviously, the more powerful the test we can come up with, the better. Given these specifications, the appropriate test to use is the **z-test** for a population mean; this is a parametric test which assumes either that the population is normally distributed and the population variance is known (the latter being as realistic as saying that 'it never rains in Wales') or that the sample is large (i.e. $n>30$). The z-test makes use of a distribution that you are already familiar with, namely the standard normal distribution; indeed, as we will see in the next section, its test statistic is directly interpretable as a standard score (see Chapters 8 and 9 to refresh your memory on z-scores). Note that the z-test is not the only test that can be used to test a hypothesis concerning a population mean (another one, based on the *t*-distribution, will be introduced in Chapter 11); nevertheless, it is the best choice given the characteristics of our example data. Here's how it works.

Step 4: Identification of the Probability Distribution of the Test Statistic and Definition of the Region of Rejection

Each test generates what is known as a **test statistic**, which is a measure for expressing the results of the test. In any one application, the numerical value of the test statistic will obviously reflect the particular sample data at hand; however, the test statistic itself is a variable quantity that can assume many different values – indeed, the probability distribution of a test statistic indicates *all* values that the statistic can assume when H_0 is true. Thus the distribution of the test statistic allows us to assess sampling error. Clearly, for a test statistic to be of any use, its probability distribution should be calculable; otherwise, we could not 'partition' our test results into those that would qualify as being significant and those that would be non-significant (see Step 2).

In many statistical tests, the probability distribution of the test statistic takes known forms (e.g. a normal distribution). In other tests, the probability distribution of the test statistic *approximates* known forms. Finally, in those instances where the probability distribution needs to be re-calculated every time the test is applied, modern computer packages will do the hard work for you, so don't panic! In Chapters 11–13, you will get to know intimately several test statistics as you go through the various exciting tests we have carefully selected for you. Here, we will simply illustrate with our example the role that the test statistic plays in the hypothesis-testing process.

The test statistic associated with a *z*-test for a population mean is known as the **z-statistic**, which is simply a *z*-score defined as follows:

$$z = \frac{\bar{x} - \mu_0}{s/\sqrt{n}}$$

where $\bar{x}$ = sample mean, μ_0 = hypothesized value of the population mean, s = sample standard deviation and n = sample size (strictly speaking, the population standard deviation (σ) should be in the denominator; however, since we do not know it, we 'cheat' a bit by using s instead; this is OK as our sample is large).

The *z*-test statistic follows the standard normal distribution which we discussed extensively in Chapter 8 and used in Chapter 9; recall that the properties of the standard normal distribution are known and that there are statistical tables available for using this distribution in practice. Thus, all *we* have to do is decide on the range of values of *z* that would lead to the rejection of the null hypothesis, in other words, establish the **rejection region**. All other values would indicate that the null hypothesis should be 'accepted' and, thus, define an **acceptance region** (but read Warning 10.1 for a cautious interpretation of 'acceptance' as non-rejection). The value of *z* that separates the rejection and acceptance regions is called the **critical value**. Assuming we have established the region of rejection, then if the computed value of the test statistic based on the sample data falls within the rejection region, we reject H_0 and take this as evidence in support of H_1. If, on the other hand, the computed value of the test statistic falls within the acceptance region, then we have no grounds for rejecting H_0 (and thus no support for H_1).

test statistic

z-statistic

rejection region

acceptance region

critical value

How do we decide on the rejection region? Well, it is here that our choice of significance level comes into play. Specifically, we use the significance level we set in Step 2 to identify the 'unlikely' values of z, given that our null hypothesis is true. In our example we decided that $\alpha = 0.05$, so we need to partition the distribution of the test statistic into an area covering 5% of the values (the rejection region) and another area covering the remaining 95% of the values (the acceptance region). Bearing in mind that for our specific null hypothesis, rejection implies that the mean has a *greater* value than that hypothesized, it becomes evident that the region of rejection should be located at the *right* tail of the distribution (in other words, it is large values of z that should be included in the rejection region). With these considerations in mind, we look into our standard normal tables and find that, at $z = 1.645$, we can define the region of rejection (see Figure 10.2(a)). This is our critical value, since 5% of the values of the test statistic exceed 1.645 and 95% are below 1.645; or, what amounts to the same thing, the probability that the test statistic exceeds 1.645 given that the null hypothesis is true is 0.05 (our significance level α) and the probability that the test statistic does not exceed 1.645 is 0.95 (which is equal to $1-\alpha$ in Table 10.3 earlier).

In the light of the above, we can restate our decision rule for rejecting H_0 as follows: If, assuming H_0 is true, the probability of obtaining a value of the test statistic as extreme as or more extreme than the critical value is less than or equal to α, we reject H_0. Otherwise we do not reject H_0. Since the significance level is instrumental in determining the region of rejection, the latter is also widely referred to as the **significant region**. Note that any change in the significance level (e.g. from 0.05 to 0.10) will affect the definition of the rejection region and, thus, will have an impact on whether the null hypothesis will be rejected or not. Therefore, it is not legitimate to change the level of significance retrospectively (e.g. if you see that, given your data, the null hypothesis cannot be rejected at, say, the 5% significance level but it would, say, at the 10% level, it is cheating if you report significant results at the more 'liberal' level).

For reasons already discussed, the region of rejection in our example was established in the right tail of the distribution of the test statistic. If our null hypothesis had been of the form $H_0: \mu \geq \mu_0$ then the region of rejection would have been established in the left tail of the distribution (see Figure 10.2(b)); however, the rationale for and determination of the relevant region would be identical to those just described (the only difference being that the 'extreme' values of the test statistic in the rejection region would be small rather than large).

What if, instead of a directional hypothesis, we have an exploratory hypothesis, leading to a null hypothesis of the form $H_0: \mu = \mu_0$? Clearly, in this instance, the extreme values of the test statistic calling for the rejection of the null hypothesis could be large or small – so where's the region of rejection? In *both* tails, of course! What we do is the following little trick: we take our significance level and chop it in half. Then if, say, $\alpha = 0.05$, we define *two* regions of rejection each capturing 0.025 of the area under the curve (i.e. each rejection area is equal to $\alpha/2$). This is shown in Figure 10.2(c). Note that we now have two critical values ($z = -1.96$ and $z = 1.96$, respectively, as our standard normal tables show) which are

significant region

WARNING 10.4

Redefining the region of rejection, in the light of the sample data is not a legitimate practice.

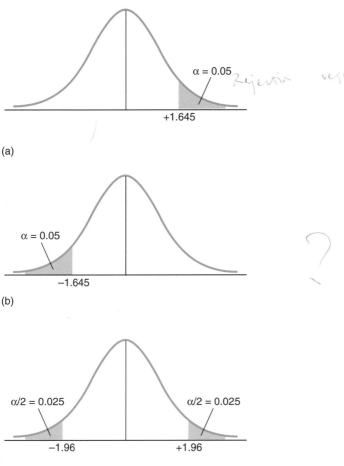

Figure 10.2   The region of rejection.

(a)

(b)

(c)

symmetrically placed and enclose between them the acceptance region (i.e. 1–α).

By convention, whenever a statistical test is used with the region of rejection defined in one tail of the test distribution only, we speak of a **one-tailed test**; if both tails of the distribution are used in defining the region of rejection, we speak of a **two-tailed test**. A one-tailed test is appropriate when a *directional* alternative hypothesis is specified (which, as already mentioned in Step 1, implies considerable prior knowledge about the nature of the hypothesized phenomenon), while a two-tailed test is the one to use if an *exploratory* hypothesis is all that can be reasonably specified (due to absence of or conflicting prior knowledge).

From the above you should be able to work out why the way in which the null and alternative hypotheses are stated is important (see Step 1). If the alternative hypothesis is directional in nature, then only one tail of the distribution will be used to define the region of rejection and this one tail will contain all the extreme values for rejecting the null hypothesis. If, on the other hand, the alternative hypothesis is exploratory in nature, both tails will have a region of rejection, each containing half the extreme values for rejecting the null hypothesis. Consequently, *at the same level of*

one-tailed test

two-tailed test

HINT 10.4

If your alternative hypothesis is directional (i.e. includes a > or < sign), then use a one-tailed test. If it is only exploratory (i.e. includes a ≠ sign), use a two-tailed test.

significance, a one-tailed test will result in the rejection of the null hypothesis more often than a two-tailed test, that is, it is *always easier to reject the null hypothesis with a one-tailed test.* This becomes immediately obvious if you compare the critical values in Figure 10.2(a) and 10.2(c); although the significance level is 0.05 in both cases, the critical value of z in Figure 10.2(a) is 1.645, while in 10.2(c) is 1.96. Thus if z turns out to be, say, 1.853 the null hypothesis would be rejected in the former case but not in the latter (a similar conclusion is reached if Figures 10.2(b) and 10.2(c) are compared). The corollary of this is that if a two-tailed test gives a significant result, then its corresponding one-tailed test will also produce a significant result (but the reverse does not always hold). Note that changing a hypothesis from exploratory to directional *after* the data have been collected (so as to justify the use of a single tail test) is not permissible; instead of being theory-driven, you become data-driven and this defeats the very purpose of hypothesis-testing.

WARNING 10.5

Reformulating a hypothesis from exploratory to directional *after* the data has been collected, is a no-no.

Step 5: Computation of the Test Statistic and Rejection or Non-Rejection of the Null Hypothesis

Having defined the region of rejection in the light of your hypothesis and chosen significance level, all that remains is to compute the value of the test statistic using the information from your sample and compare it to the critical value(s). In our wonderful example, the computed value of the test statistic is as follows (see Step 1 to remind yourself where these figures came from):

$$z = \frac{\bar{x} - \mu_0}{s/\sqrt{n}} = \frac{3.5 - 3}{2.4/\sqrt{200}} = 2.94$$

Going back to Figure 10.2(a), we can see that the critical value (at the 5% significance level) is 1.645. As our computed test statistic exceeds the critical value, it falls within the region of rejection and, therefore, we reject H_0 in favour of H_1. Thus we conclude that the average number of monthly visits to the hairdresser by teenagers is more than 3. Fame and fortune, at last!

We can also compute the probability of obtaining a test statistic as extreme as or more extreme than the one actually obtained by looking at the tables of the standard normal distribution. This probability is known *p*-value as the **p-value** and, the lower it is, the stronger is the evidence against the null hypothesis. In our example, the p-value associated with a z-value of 2.94 is 0.0016. This shows us that, if the null hypothesis were true, we would have observed a test statistic of 2.94 (or more extreme) less than twice in one thousand times.

The p-value provides a bit more information on how far down in the significant region our result lies. Articles in scientific journals often discuss their results in terms of p-values: they are full of statements such as $p<0.05$, $p<0.01$, $p<0.001$ and so on. Unlike critical values, which are test-specific (e.g. a critical value of a z-test and a critical value of a chi-square test will be very different for a given α), p-values represent a 'common currency' across which the results of different tests can be compared. In this context, it is better practice to report the actual *p*-value (e.g. $p = 0.004$, $p = 0.013$, etc.) rather than use the conventional cut-off points (i.e. $p < 0.01$, $p < 0.05$, etc.). This allows readers to form their own judgment about the significance of the results, as different people may feel that different levels of significance are appropriate; reporting *p*-values keeps everybody happy by enabling everyone to decide individually on the strength of the evidence against the null hypothesis. Needless to say, you *must* accompany *p*-values with some information indicating whether they relate to a one- or two-tailed test (see Step 4 above); this is usually accomplished by such statements as '*p*<0.05, sum of both tails', or 'one-tailed *p*<0.001'. Failure to do so is bound to result in confusion.

The discussion of *p*-values concludes our presentation of the hypothesis-testing procedure. As we mentioned in the beginning of this chapter, some of the above steps will be executed by your statistical package. While you cannot really expect your computer to set hypotheses on your behalf (Step 1), decide on a significance level (Step 2) and choose an appropriate test (Step 3), once you have made these decisions, it will automatically know which test statistic to use and how to present significant results. In the next three chapters, you will have the once-in-a-lifetime opportunity to see several examples of computer-generated output for a variety of statistical tests carefully selected by yours truly (a quick visit to Chapter 5 to refresh your memory on statistical analysis packages would be a good idea at this stage). But we're not finished yet – not quite.

Hypothesis-Testing and Confidence Intervals

Consider a situation where you want to test an exploratory hypothesis concerning, say, a mean. For the sake of convenience, let us use our existing example of teenager visits to the hairdresser but re-formulate our initial hypotheses (see Step 1 in previous section) as follows:

$$H_0: \mu = 3$$
$$H_1: \mu \neq 3$$

Is there a way of testing H_0 at a significance level α without applying a significance test? Yes, there is – and you already know how to do it (even if you don't know that you know). Instead of following the hypothesis-testing procedure just described, let us construct a $100(1-\alpha)\%$ confidence interval for μ. Then if 3 is included in this interval, we do not reject H_0; if, on the other hand, 3 is not included in the interval, we reject H_0 in favour of H_1. So, if we set $\alpha = 0.05$, we need to set a 95% interval for μ using our sample data. Following the procedures so eloquently presented in Chapter 9, the 95% confidence interval in our case is as follows (all the strange symbols are as defined in Chapter 9):

$$\bar{x} - k_c s_{\bar{x}} \leq \mu \leq \bar{x} + k_c s_{\bar{x}} = \bar{x} - k_c(s/\sqrt{n}) \leq \mu \leq \bar{x} + k_c(s/\sqrt{n})$$

Placing the appropriate values into this equation results in the following calculations:

$$= 3.5-1.96(2.4/\sqrt{200}) \leq \mu \leq 3.5+1.96(2.4/\sqrt{200})$$
$$= 3.17 \leq \mu \leq 3.383$$

Since the 95% confidence interval for μ does not include 3, we reject the null hypothesis and conclude that the population mean is not 3; indeed, from the upper and lower levels of the confidence interval it can be seen that the true mean is likely to be greater than 3.

Would we have arrived at the same conclusion had we applied a two-tailed z-test for a population mean? Indeed, we would. The z-value from our data comes to 2.94 (see previous section), while the critical values for a two-tailed z-test and a 5% significance level come to -1.96 and 1.96, respectively (see Figure 10.2(c)). As the calculated value of the test statistic lies outside the critical values, we reject H_0; moreover, since the relevant region of rejection turns out to be in the right tail of the distribution, we can conclude that the true population mean is likely to be higher than the hypothesized mean.

In general, hypotheses involving population parameters can be tested either by following the hypothesis-testing procedure outlined previously or by setting a confidence interval and observing whether the hypothesized value of the population parameter falls within or outside this interval. Given a significance level α, the corresponding confidence interval is $100(1-\alpha)$; recalling that $1-\alpha$ corresponds to the confidence coefficient associated with interval estimation (see Chapter 9), we can see that the significance level is the complement of the confidence coefficient and vice versa. In this context, the confidence interval approach to hypothesis-testing focuses on the region of acceptance, while the significance approach to hypothesis-testing focuses on the region(s) of rejection; both produce identical results.

HINT 10.5

For hypotheses involving population parameters, significance testing and confidence interval estimation are equivalent procedures.

Statistical and Substantive Significance

Before we say goodbye to this incredibly exciting chapter, a few words of caution are in order.

First, newcomers to hypothesis-testing often confuse a significant result with an *important* result. The fact that your test produced a significant value does not necessarily mean that you have an important or even remotely interesting finding. For example, if you compare the average weight of 20 elephants with the average weight of 20 ducks, surprise, surprise, you are going to get a significant result (probably at $p<0.00000001$!); however, reporting a significant result for a hypothesis so blatant is not going to impress many people. Thus, if your hypotheses are trivial, significant findings are not going to cut much ice.

Second, the fact that you have observed a significant result (even with a reasonable hypothesis), tells you nothing about the *magnitude* of the effect involved. For example, observing a significant difference between the average annual income of Greek and German cost accountants does not necessarily mean that the difference is of *practical* significance (the difference may be very small – after all, if the former earn £2 a year more than the latter, does this make them richer?). Recall that the power of a statistical test increases with the sample size; thus, even a tiny difference may turn out to be statistically significant if very large samples are involved. Indeed, with very large sample sizes, many statistical tests become extremely sensitive and will declare as significant even minor deviations from the null hypothesis. Yes, we know that we told you in Chapters 2, 6 and 9 that you should try and work with large samples. However, as with everything in life, (very) large samples have also their downside, part of which is that statistical power may become *too* high. You can never win, can you?

Third, do not be tempted to dismiss your findings because they turn out to be *non*-significant. Granted, often a null hypothesis is formulated with the express purpose of rejecting it as the real interest lies in the alternative hypothesis (i.e. you want to find differences, establish relationships, etc.). However, failure to reject the null hypothesis can be just as interesting/ important/exciting as rejecting it or even more so. For example, if you are testing the well-established theory that penguins are much more introverted than salamanders and are unable to reject the null hypothesis, your finding would contain a high degree of surprise; consequently, it would probably generate more discussion among the scientific community than if you had simply supported conventional wisdom one more time. After all, finding that penguins are *not* more introverted than salamanders (when everybody thought they were) is bound to hit the front pages! So, a non-significant finding is not necessarily a bad finding and, conversely, a significant finding is not necessarily a good finding.

Fourth, and related to the above point, sometimes the aim is to 'fail' intentionally, i.e. *not* to reject the null hypothesis. For example, if you are applying a statistical test to check the distribution of your variable against the normal distribution, you want to get non-significant results because this would indicate that making the assumption of normality would indeed be justified (we will see an example of such a test in Chapter 11). Under these circumstances, you are deliberately interested in obtaining **null results**, that is findings that would not lead to the rejection of the null hypothesis.

null results

Fifth, do not be tempted to use your computer capabilities to run test after test on the same set of data in a mad search for significance. Resist this temptation because

> any large set of data – even several pages of a table of random numbers – contains some unusual pattern. Sufficient computer time will discover that pattern, and when you test specifically for the pattern that turned up, the test will be significant. It will also mean exactly nothing.[6]

Sixth, do not confuse the process of *discovery* with the process of *verification*. It is perfectly legitimate, as a result of exploratory hypothesis-testing, to come up with more specific (i.e. directional) hypotheses concerning the topic at hand. However, it is not legitimate to test these hypotheses on the same set of data that suggested them in the first place. What you discovered may simply be an artefact of the specific data set you happen to have and not be replicated in other samples. Consequently, you should get hold of another set of data in order to verify that your hypothesis developed previously is in fact reasonable. Yes, this means extra work, but who told you that the way to scientific stardom is paved with roses?

Taken together, the above comments suggest that statistical significance should never be allowed to obscure or overrule **substantive significance**. At the end of the day, it is the substance of your findings in terms of their implications for theory, practice or policy that matters, not whether these findings happen to be significant or not. Significance tests are only tools that help us deal with issues of theoretical and/or practical relevance and, in themselves, have no intrinsic value whatsoever. You should always remember this.

substantive significance

WARNING 10.7

Never allow significance testing to become an end in itself.

Summary

The purpose of this chapter was to introduce you to the second approach to statistical inference, namely hypothesis-testing. First, we looked at hypotheses from a number of different perspectives and highlighted the indirect nature of obtaining evidence in support of a hypothesis. We then moved on to consider the practicalities of hypothesis-testing by going through a five-step procedure which can be used as a blueprint for testing any hypothesis. Following this, we examined the relationship between hypothesis-testing and interval estimation and concluded this chapter with some words of wisdom concerning the role and interpretation of statistical significance. If you have grasped the issues in this chapter, then you should be able to take Chapters 11–13 in your stride; if you are still a bit uncertain, go through this chapter once more – perhaps a bit slower this time.

Questions and Problems

1. What is a hypothesis?

2. Give an example of a null hypothesis and its corresponding alternative hypothesis.

3. What are the five steps in hypothesis-testing?

4. Under what circumstances would you formulate directional hypotheses?

5. What are the two types of error associated with hypothesis testing? Which one is more important?

6. Explain what the terms 'significance level' and 'region of rejection' mean.

7. What are the main considerations in selecting an appropriate significance test?

8. When would you use a one-tailed versus two-tailed test? Use an example to illustrate your answer.

9. Why should statistical significance be distinguished from substantive significance?

10. How can you use your knowledge of hypothesis-testing to impress your least-liked relative?

Notes

1. Daniel, WW & Terrell, JC (1995) *Business Statistics for Management and Economics*, 7th edn, p. 322. Boston: Houghton Mifflin.
2. Kanji, GK (1993) *100 Statistical Tests*, p. 1. London: Sage.
3. Blalock, HM Jr (1979) *Social Statistics*, revised 2nd edn, p. 105. Tokyo: McGraw-Hill Kogakusha.
4. Daniel, WW & Terrell, JC (1995) *Business Statistics for Management and Economics*, 7th edn, p. 234. Boston: Houghton Mifflin.
5. Moore, DS (1979) *Statistics: Concepts and Controversies*, p. 286 and pp. 287–288. San Fransisco: WH Freeman.
6. Moore, DS (1979) *Statistics: Concepts and Controversies*, p. 294. San Francisco: WH Freeman.

Further Reading

Henkel, RE (1976) *Tests of Significance*. London: Sage. A good introduction to the principles of hypothesis-testing, highlighting key concepts and techniques; the issues relating to the interpretation of significance tests are particularly worth noting.

Kanji, GK (1993) *100 Statistical Tests*. London: Sage. The title speaks for itself. This is a great reference book which in some 200 pages manages to cram just about every test you are ever going to need (plus a few more!); excellent for choosing the correct type of test.

Mohr, LB (1990) *Understanding Significance Testing*. London: Sage. A concise treatment of hypothesis-testing, providing a complementary perspective to that of Henkel (1976) above.

11

Simple things first:
one variable, *one* sample

Single Sample Hypotheses

In this chapter, we start applying the basic ideas of hypothesis-testing that you learned in Chapter 10 to specific problems (otherwise you would bitterly complain that you get too much theory and too little practice!). As you may be a bit mentally stressed from Chapters 9 and 10, we are going to take things easy here and leave the more tricky stuff for later. Specifically, we are going to look at formulating and testing hypotheses relating to one variable at a time as applied to a single sample; such hypotheses focus upon the characteristics of a single population. After you have mastered these, we will show you how to make comparisons between different groups/measures (Chapter 12) and how to investigate relationships between variables (Chapter 13).

So what kinds of hypotheses can we develop for a single population? Well, quite a few actually.

For starters, we can set up a hypothesis concerning the *distribution* of the variable; for example, we may hypothesize that our variable follows a normal distribution in the population and then use our sample data to test whether this is likely to be the case. Statistical tests aimed at determining the extent to which the distribution of a variable follows some pre-specified

goodness-of-fit tests functional form in the population are known as **goodness-of-fit tests**; they compare the extent to which the observed (i.e. empirical) frequencies 'fit' the expected (i.e. theoretical) frequencies.

Another hypothesis we can set up relates to the *central tendency* of the variable; for example, we may hypothesize that the mean of our variable has a certain value in the population and then use our sample information to test whether this is likely to hold true. Statistical tests aimed at determining whether a population mean or median takes on a particular value

tests for location are known as **tests for location**; the example used throughout Chapter 10 to illustrate the nature of hypothesis-testing utilized such a test (namely the z-test for a population mean).

Third, we can set up hypotheses concerning the degree of *dispersion* in a variable; for example, we may hypothesize that the variance of our variable has a certain value in the population and then test for this using our sample data. Statistical tests aimed at testing whether a population

tests for variability variance is of a certain magnitude are known as **tests for variability**.

Fourth, with dichotomous variables (if you have forgotten what these look like, have a quick peek at Chapter 3), we can set up hypotheses relating to *proportions* (or percentages); for example, we may hypothesize that only π per cent of the population has a certain characteristic (while $100-\pi$ does not) and then test this hypothesis using our sample data. Tests of this type are commonly known as **tests for proportions**.

tests for proportions

Lastly, it would only increase your hatred for statistical testing, if we told you that there are also **tests for randomness** which, among other things, can be used to check that a given sample is indeed a random one. So, we won't – only kidding, we'll tell you all about them later in the chapter.

tests for randomness

Figure 11.1 summarizes the above discussion by providing an overview of the various population characteristics for which we can formulate hypotheses and associated techniques for testing them. You should use Figure 11.1 as a 'navigation chart' each time you want to test a hypothesis concerning a single population; similar charts will be provided free of charge in Chapters 12 and 13.

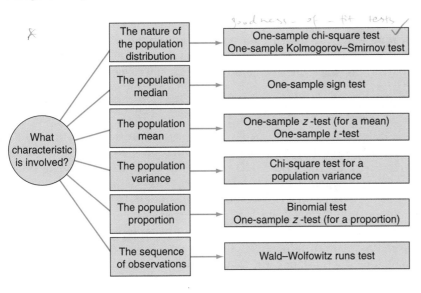

Figure 11.1   Statistical techniques for hypotheses involving population characteristics.

Assessing Fit

At an intuitive level, the notion of 'fit' is perhaps grasped best by trying to imagine a situation where there is an 'ideal' scenario and an 'actual' turn of events. Say, for example, that you are a final-year student of Spanish mediaeval ecclesiastical history and are considering your lecture program. Now, you know that your total weekly lecture load is going to be 60 hours (only) but you do not yet know when these lectures will be scheduled. Let us assume that you would ideally like to spread your lecture program evenly through the week (i.e. attend a mere 12 hours of lectures, every day, Monday to Friday inclusive!). When you actually get your timetable, you find out that you have 12 hours of lectures on Monday, 10 on Tuesday, 14 on Wednesday, 6 on Thursday and 18 on Friday. How close is the actual timetable to your ideal one? In other words, what is the *fit* between what you had wished/prayed for and what you got? Would a timetable with 12 hours Monday to Wednesday, none on Thursday and 24 hours on Friday (sorry, no sleep) represent a better or worse fit?

one-sample chi-square test, one-sample Kolmogorov–Smirnov test

It is questions of this kind that we try to answer when we use goodness-of-fit tests. We will consider two of these tests here: the **one-sample chi-square test** and the **one-sample Kolmogorov–Smirnov (K-S) test** (what a mouthful!). The former is based on a consideration of *absolute* frequencies and, thus, it can be applied to any variable irrespective of the level of measurement (although it is typically used with nominal data). In contrast, the one-sample K-S test utilizes *cumulative* frequencies and, thus, implicitly assumes that the variable concerned is, at least, ordinal (see Chapter 7 if you cannot remember the distinction between absolute and cumulative frequencies).

The One-Sample Chi-Square Test

observed frequencies theoretical frequencies

This the test to go for whenever you want to compare a set of **observed frequencies** with a set of **theoretical frequencies**. By observed frequencies, we mean frequencies calculated from the empirical data (see Chapter 7), reflecting the *actual* distribution of the variable concerned in the sample (hence they are also known as 'actual frequencies'). By theoretical frequencies we mean frequencies generated on the basis of prior knowledge or theoretical considerations, reflecting our expectations regarding the distribution of the variable in the population (hence they are also known as 'expected frequencies'). The question then arises whether the differences between observed and theoretical frequencies are significant.

The null hypothesis under the chi-square one-sample test is that no difference exists between observed and theoretical frequencies (or, what amounts to the same thing, that the observed frequencies are equal to the theoretical frequencies). If the observed frequencies depart *significantly* from the theoretical frequencies (i.e. cannot be reasonably explained by sampling fluctuations), we take this as evidence for rejection of the theory/speculation that gave rise to the theoretical frequencies. If, on the other hand, we find that the differences between observed and expected frequencies are small and non-significant, then we have no grounds for rejecting the theory. Note that a 'reverse' hypothesis-testing logic characterizes the one-sample chi-square test: our interest is in obtaining null results rather than finding support for the alternative hypothesis via a significant finding (on this point, see the discussion on substantive significance in Chapter 10).

Okay, we know that you are dying for an example, so here is one: Say that you want to test TV viewers' preferences for 'late-nite' horror movies on the following four British TV channels: BBC1, BBC2, ITV and Channel 4. Let us assume that you have no reason to believe that one channel would be preferred to another, since all appear to provide an excellent selection of gory stuff. If this is indeed the case, you would expect equal preferences, that is approximately the same number of people would prefer BBC1, BBC2, etc.; put in disgustingly complicated statistical jargon, your null hypothesis is that viewing preferences in the population follow a *uniform* distribution (see Chapter 8). You now collect some data on the viewing preferences of a random sample of 111 individuals and code them on the variable TVHORROR as follows: 1 = BBC1, 2 = BBC2, 3 = ITV and 4 = Channel 4. Given that you hypothesize equal preferences, you would expect approximately 28 individuals to select each channel as being 'the best' (i.e. 111 individuals spread equally over 4 channels or 111/4 = 27.75). However, what you observe is that 11 sample respondents prefer BBC1, 64 BBC2, 16 ITV and 20 Channel 4. To test whether these frequencies are significantly different from your expectations, you run a one-sample chi-square test using your trusted computer; what you should get is something like Table 11.1 (the latter was produced by the *SPSS/PC+* data analysis package discussed in Chapter 5; other packages should produce similar output).

TVHORROR			
		Cases	
Category	Observed	Expected	Residual
1.00	11	27.75	−16.75
2.00	64	27.75	36.25
3.00	16	27.75	−11.75
4.00	20	27.75	−7.75
	Total 111		
Chi-Square	D.F.	Significance	
64.604	3	0.000	

Table 11.1 An example of the one-sample chi-square test

According to the results, the hypothesis of equal viewing preferences is not supported, as the test statistic (chi-square = 64.604) is highly significant ($p<0.000$). Since our statistical program prints only three places after the decimal point, we cannot see exactly how (highly) significant this result is. A safe bet would be 0.001; but who knows, it may be 0.000 000 001. If you really need to find out – and usually this should not be necessary – you need to tease more decimal places out of your program. In any event, going back to our example, the highly significant result is not surprising given the large discrepancies between the 'Observed' and 'Expected' values as indicated in the 'Residuals' column. In this context, more people preferred BBC2 than expected (positive residual) and fewer people than expected preferred BBC1, ITV and Channel 4 (negative residuals); in absolute terms, the largest discrepancy is noted with BBC2 (36.25) and the smallest with Channel 4 (7.75). Overall, viewing preferences across TV channels are not equal.

In the above example, the expected frequencies were all specified to be the same. This is by no means a requirement of the chi-square test and any theoretical (expected) frequencies can be specified and compared to observed (actual) frequencies. To illustrate this point, Table 11.2 shows the results of a one-sample chi-square test applied to the same data, but under the null hypothesis that BBC2 is preferred by thirteen times as many people as any of the other channels (note that, according to the results, this hypothesis must also be rejected as the theoretical frequencies do not 'fit' the data well). A summary of the steps involved in applying the one-sample chi-square test is generously provided in Table 11.3.

Table 11.2 Another example of the one-sample chi-square test

TVHORROR			
		Cases	
Category	Observed	Expected	Residual
1.00	64	7.00	57.00
2.00	11	90.00	−79.00
3.00	16	7.00	9.00
4.00	20	7.00	13.00
	Total 111		

Chi-Square	D.F.	Significance
569.202	3	0.000

Table 11.3 Applying the one-sample chi-square test

1. Specification of the categories of the variable of interest.
2. Determination of the expected (theoretical) frequencies for each category (based on theory, prior research, etc.).
3. Determination of the observed (actual) frequencies for each category.
4. Comparison of observed versus expected frequencies and calculation of chi-square statistic.
5. Examination of significance of chi-square statistic and rejection (or non-rejection) of null hypothesis.

When the number of categories in your variable is greater than 2 (as in our example above), you should not really use the chi-square test if (a) more than 20% of the expected frequencies are smaller than 5, or (b) if any expected frequency is less than 1 (don't ask us why; as with all tests, there are some silly conditions that have to be met even if this means that they make life complicated). If you face any of the above problems (not an unusual situation, particularly with small samples and/or too many categories), then try to combine *adjacent* categories (assuming of course that such combinations are meaningful). This usually sorts things out. Another thing to watch out for is that if you only have two categories (i.e. you are dealing with a dichotomous variable), then both expected frequencies must be 5 or larger; if they are not, combining categories is obviously not an option (as you would end up with a single category) and you have to use another test (such as the binomial test, discussed later in this chapter).

Before we part company with this illustrious test, you should note that the one-sample chi-square test is often used to examine the hypothesis that a variable follows a normal distribution (or, more precisely, that the sample data has been drawn from a population which is normally distributed). While this is a frequent application, 'the general opinion is that the K-S test of goodness of fit is more powerful than the chi-square test used for the

WARNING 11.1

Check that your expected frequencies meet the minimum levels required by the one-sample chi-square test before applying it to your data.

same purpose';[1] bearing this in mind and given that we will be discussing the K-S test next, we see little benefit in demonstrating the use of chi-square for normality testing purposes (check out the Further Reading section if you are desperate to find out).

The One-Sample Kolmogorov–Smirnov (K-S) Test

HINT 11.1

Use the one-sample K-S test to check whether your variables are likely to be normally distributed in the population *before* applying statistical procedures that assume normality of distribution.

This tongue twister is *the* test to use in order to check that the distribution of a variable follows a particular form. By far, its most common application is in testing whether observed values can reasonably be thought to have come from a normally distributed population (although it can be used to test for other distribution types, such as uniform or Poisson). In this context, the one-sample K-S test is an excellent check to apply *before* using statistical procedures that rest on the assumption of normality (see Chapters 6 and 10 and also the discussion on the one-sample *t*-test later in this chapter). By doing this you will know whether the assumption of normality is tenable for the variable in question (thus pre-empting vicious criticism from statistical purists who desperately try to catch you out!).

The null hypothesis under the one-sample K-S test is that no difference exists between the observed distribution and a specified theoretical distribution (e.g. normal). The test computes the cumulative relative frequencies that would occur under the theoretical distribution and compares them with the observed cumulative relative frequencies; it then determines the point at which the two sets of frequencies diverge the most, i.e. 'the maximum deviation' between the cumulative theoretical frequency distribution and the cumulative observed frequency distribution. (See Chapter 7 for a memory refreshment on cumulative frequency distributions.) If this maximum deviation is significant (i.e. cannot be reasonably explained by sampling fluctuations), there is evidence that the observed distribution does not follow the specified form. If, on the other hand, the maximum deviation is not significant, there is no reason to believe that the observed distribution departs from the theoretical specification. As the one-sample K-S test is concerned with the degree of agreement between the distribution of a set of observed scores and some specified theoretical distribution, it is also a goodness-of-fit test (as is the one-sample chi-square test).

Let us continue using our example on TV viewing habits to illustrate the application of the one-sample K-S test. Suppose that, in addition to TV channel viewing preferences, you also obtained data from your random sample of 111 individuals on the number of horror movie videos that they rented over the past year. Now you want to test whether this variable (let's call it VIDEOHOR) is normally distributed in the population. You ask your computer to perform a one-sample K-S test and you get the results shown in Table 11.4 (again, this is *SPSS/PC+* output, but any decent package should be able to do a one-sample K-S test; if not, throw it away!).

The first thing to note from Table 11.4 is that, in this example, you are comparing the distribution of your variable against a normal distribution which has the same mean (114.31) and the same standard deviation (83.94) as your sample distribution. This is the **best-fitting normal distribution** for your data since no other normal curve would provide a

best-fitting normal distribution

Table 11.4   An example of the one-sample Kolmogorov–Smirnov test

VIDEOHOR					
Test distribution – Normal				Mean: Standard deviation:	114.31 83.94
Cases: 111					
	Most Extreme Differences				
Absolute	Positive	Negative		K-S *z*	2-tailed *p*
0.13166	0.13166	−0.09242		1.381	0.044

better fit (indeed, the best-fitting normal distribution for any observed frequency distribution is the one having the same mean and the same standard deviation as those computed from the observed distribution).

The second thing to note is that the test statistic (K-S $z = 1.381$) is significant ($p<0.05$), indicating that the absolute 'Most Extreme Difference' (i.e. the maximum deviation) is too large to have come about if the distribution was normal. Thus, according to the results, the number of horror videos rented over the past year does not follow a normal distribution.

Note that you do not *have* to test for normality of distribution by using the best-fitting normal distribution (as done above). You can specify the parameters of the theoretical distribution and then test your observed distribution against the theoretical distribution with the specified parameters. To appreciate this application, imagine that somebody else had undertaken a similar study in the past and concluded that the average number of horror videos rented was 95 with a standard deviation of 66. You are dying to prove him wrong, so you apply the K-S one sample test to *your* data specifying a normal distribution with *his* parameters; the results are shown in Table 11.5. Not surprisingly, the fit of this distribution is also poor and, as an inspection of the absolute 'Most Extreme Difference' against that in Table 11.4 shows, considerably worse than that obtained using the best-fitting normal distribution as the theoretical model. The steps involved in applying the one-sample K-S test are summarized in Table 11.6.

Table 11.5   Another example of the one-sample Kolmogorov–Smirnov test

VIDEOHOR					
Test distribution – Normal				Mean: Standard deviation:	95.00 66.00
Cases: 111					
	Most Extreme Differences				
Absolute	Positive	Negative		K-S *z*	2-tailed *p*
0.15706	0.08437	−0.15706		1.647	0.009

1. Specification of the theoretical distribution of interest.
2. Determination of the cumulative relative frequencies for the theoretical distribution.
3. Determination of the cumulative relative frequencies for the observed distribution.
4. Comparison of the two sets of frequencies and calculation of maximum absolute deviation.
5. Examination of significance of maximum deviation and rejection (or non-rejection) of null hypothesis.

Table 11.6 Applying the one-sample Kolmogorov–Smirnov test

Testing for Location

You will recall from Chapter 8 that one aspect of summarizing data involves the calculation of 'average' or 'typical' values, such as means and medians, which describe the *central location* in a variable. In many situations, we are less interested in the average values we obtain from our sample data and more concerned with the corresponding population parameters. Thus in the video rental example discussed earlier, what may be of particular importance to, say, a horror film producer, is not the average number of horror videos rented by your specific sample of 111 individuals but the average number of videos rented by the UK population as a whole. Now, as you know from Chapter 9, one thing you can do is estimate the latter from your sample data by setting up a confidence interval. Another thing you can do is test whether the population parameter (i.e. mean or median in this case) takes on a particular value; for example, you may want to test the hypothesis that, on average, UK residents rent 100 horror videos a year. Or you may want to test the hypothesis that, on average, the number of horror videos rented is greater (or less) than 300.

It is hypotheses like these (not necessarily constrained to the rental of horror videos) that location tests are designed for. Table 11.7 provides a general overview of the various hypotheses that can be tested with location tests. We will consider two of the most useful tests here, the **one-sample sign test** and the **one-sample *t*-test**; the former requires data that are at least ordinal in nature, whereas the latter assumes at least interval-level measurement.

one-sample sign test
one-sample *t*-test

Null hypothesis	Alternative hypothesis	Type of test
$M = M_0$	$M \neq M_0$	2-tailed
$M \leq M_0$	$M > M_0$	1-tailed
$M \geq M_0$	$M < M_0$	1-tailed

M = population mean (μ) or median (m); M_0=hypothesized value.

Table 11.7 Hypotheses for one-sample location tests[a]

The One-Sample Sign Test

If you have a hypothesis about a *median*, this test should do the trick. Let us assume that you want to find out whether a median, *m*, takes a certain value in the population (i.e. your null hypothesis is that $m = m_0$). Remember that the median splits the distribution into two equal parts (see Chapter 8), so if the null hypothesis is correct about one-half of the observations should be larger than the specified median value and about one-half should be smaller than the median. What the sign test does is to (a) replace each observation less than m_0 with a minus (–) sign, (b) replace each observation greater than m_0 with a plus (+) sign, and (c) count the number of plus and minus signs (while ignoring zero differences). It then determines whether so many (or so few) plus signs could have come about given a median value of m_0. If yes, then we have no grounds for rejecting the null hypothesis (and we can go to the pub to celebrate); if not, then we have to conclude that the population median is not equal to the hypothesized value, m_0 (and we can go to the pub to drown our sorrows!).

For no other reason but to make life complicated, the test statistic of the one-sample sign test depends on the size of the sample (yes, we agree that whoever discovered this should be locked away, but that's the way it is). With small samples ($n < 30$), the test uses the binomial distribution and provides an exact probability value (see also Chapter 9 and the section on testing proportions later in this chapter). With larger samples ($n > 30$), the normal approximation to the binomial is used, which gives a z-value and an associated probability. As we are perfectionists of the highest possible standards, we shall illustrate both versions of the test using our celebrated example of horror videos.

Let us assume that you want to test the hypothesis that the median number of horror videos rented over the past year is 163; you picked this number while reading the *Transylvanian Times* arts page, which reported on a study of horror video rentals in Romania (and, having nothing better to do, you want to see whether UK residents have similar habits). Let us further assume that you have randomly split your total sample of 111 individuals into one small sub-sample of, say, 27 respondents and a larger sub-sample containing the rest. Tables 11.8 and 11.9 show the output obtained from testing the hypothesis $m = 163$ on the small and large sub-samples, respectively (surprise, surprise, we are again using *SPSS/PC+* – we just *love* this package!).

Table 11.8 An example of the one-sample sign test: small sample

MEDVAL with VIDEOHOR			
Cases			
18	– Diffs	(VIDEOHOR Lt MEDVAL)	
9	+ Diffs	(VIDEOHOR Gt MEDVAL)	(Binomial)
0	Ties		2-tailed $p = 0.1237$
—			
27	Total		

Other than the difference in the test statistic, the output of the test is identical for the small and the large sub-samples, indicating the number of positive and negative differences (i.e. plus and minus signs) and the number of ties (here equalling zero). MEDVAL is a constant which has been set to 163, consistent with the hypothesis that the median of VIDEOHOR in the population takes this value. Note that, as far as the small sub-sample is concerned, we cannot reject the null hypothesis that $m = 163$ since the

MEDVAL
with VIDEOHOR

Cases

61	− Diffs	(VIDEOHOR Lt MEDVAL)	
23	+ Diffs	(VIDEOHOR Gt MEDVAL)	$z = 3.9515$
0	Ties		2-tailed $p = 0.0001$
—			
84	Total		

Table 11.9 An example of the one-sample sign test: large sample

test is non-significant ($p>0.10$). In contrast, for the larger sub-sample, the test turns out to be highly significant ($p<0.000$), implying that the median number of rented videos by UK individuals is not 163; in fact, as there are a lot more minus signs, it seems that the population median is less than 163. Note that if we were to test this directional hypothesis (i.e. that $m<163$) on the smaller sub-sample, we would also obtain a significant result (at the 10% significance level), since the corresponding one-tailed probability would come to $0.1237/2 = 0.06187$ (see Chapter 10 for the calculation of one-tailed versus two-tailed significance levels). The various steps involved in applying the one-sample sign test are summarized in Table 11.10.

1. Specification of the median value and formulation of the desired null hypothesis (see Table 11.7).
2. Comparison of the median value with the individual observations.
3. Replacement of the differences with plus and minus signs and count of the number of positive and negative signs.
4. Examination of binomial exact probability (small samples) or significance of z-statistic (large samples) and rejection (or non-rejection) of appropriate null hypothesis.

Table 11.10 Applying the one-sample sign test

As a final point, it is worth mentioning that the one-sample sign test can also be applied to test hypotheses concerning the population mean (assuming that the distribution of the latter is roughly symmetrical). The procedure is identical to the one discussed above, the only difference being that the plus and minus signs reflect departures from the hypothesized mean rather than median. Although there are other more powerful tests for testing hypotheses involving means (such as the one-sample t-test discussed next), the one-sample sign test is worth a try when the sample is small *and* the distribution in the population is non-normal (i.e. when you are truly desperate!).

The One-Sample *t*-Test

If you want to test a hypothesis concerning a mean, you cannot go wrong if you use the one-sample t-test. Say that you want to test whether the population mean exceeds a certain value (i.e. that $\mu > \mu_0$ which, in null hypothesis terms, implies that $H_0 : \mu \leq \mu_0$ – see Table 11.7). The one-sample t-test (a) computes the difference between the sample mean and the hypothesized value (i.e. $\bar{x}-\mu_0$), (b) takes into account the sample size, n, as well as the likely variability in the population (using the standard deviation of the sample, s, as an estimate), and (c) determines whether the sample is likely to have come from a population whose mean value exceeds μ_0. If yes, this is reflected in a significant one-tailed test statistic (t-value) and we can reject the null hypothesis; if, on the other hand, the t-value is non-significant, we have no grounds to conclude that the population mean is indeed greater than the pre-specified value, μ_0.

Let us stick with our tried and tested example of horror videos and use the one-sample t-test to test the hypothesis that, on average, UK consumers rent more than 100 horror videos annually. Our variable is again VIDEOHOR and the output obtained from the *SPSS/PC+* package (which did all the hard work) is gracefully displayed in Table 11.11.

Table 11.11 An example of the one-sample *t*-test

VIDEOHOR
MEANVAL

Variable	Number of cases	Mean	Standard deviation	Standard error
VIDEOHOR	111	114.3091	83.943	8.004
MEANVAL	111	100.0000	0.000	0.000

(Difference) Mean	Standard deviation	Standard error	t-value	Degrees of freedom	2-tail probability
14.3091	83.943	8.004	1.79	110	0.077

You get quite a bit of information from a typical t-test output. This includes the sample size, the sample mean, the sample standard deviation and, as a bonus, the standard error of the mean (go back to Chapter 9 if you are not entirely sure what the standard error is). MEANVAL is not a variable but a constant set to the hypothesized mean population value (hence, not surprisingly, its standard deviation and standard error are both zero). Next, the 'Mean Difference' is displayed (which is the difference between the sample mean and the pre-specified value, i.e. 114.3091–100) and its associated standard deviation and standard error; not surprisingly, the latter two statistics are identical to the sample standard deviation and standard error, as there is only one sample involved. Finally, the value of the t-statistic is shown together with a two-tailed level of significance (this is the default in the *SPSS/PC+* package). Note that the t-value is simply computed by dividing the 'Mean Difference' by the standard error (i.e. $t = 14.309/8.004 = 1.79$), while the degrees of freedom are equal to $n-1$, where n = sample size (see Chapter 9).

So what can we make out of all this? First, as we are testing a directional hypothesis we need to halve the p-value (see Chapter 10). This comes to $0.077/2 = 0.0385$, which indicates a (one-tailed) significant result (at $p<0.05$). However, before we can proudly announce that the mean number of horror videos rented in the population exceeds 100, we must look at the direction of the mean difference (to ensure that we are at the 'correct' tail of the t-distribution, in this case, to the right of the hypothesized mean). Here the mean difference is indeed in the hypothesized direction and, therefore, we find support for the hypothesis that the mean number of horror videos rented annually exceeds 100. A summary of the procedure associated with applying the one-sample t-test is given in Table 11.12.

1. Specification of the population mean value and formulation of the desired null hypothesis (see Table 11.7).
2. Calculation of the sample mean value and associated standard error.
3. Calculation of the difference between the sample mean value and the hypothesized population value and computation of t-statistic.
4. Examination of significance of t-statistic and rejection (or non-rejection) of appropriate null hypothesis.

Table 11.12   Applying the one-sample t-test

WARNING 11.2

When testing directional hypotheses and using one-tailed significance tests, make sure that any significant result is in the correct direction. Otherwise you may support exactly the *opposite* of what you hypothesized!

At this stage you are no doubt wondering why we have been using the one-sample t-test to test a hypothesis concerning a mean, while in Chapter 10 we used the one-sample z-test for exactly the same purpose. The reason for this is not a deliberate attempt to confuse you, but the fact that statisticians, in their infinite wisdom, recommend that different tests are appropriate, depending upon (a) whether the underlying population is normally distributed or not, (b) whether the sample size is large or small, and (c) whether the variance of the population is known or unknown. Thus, if we wanted to be 'statistically correct', we should consider no fewer than eight possibilities simply in order to decide how to test a hypothesis concerning a single mean.

In the interests of your sanity (and ours), we decided to take a more pragmatic approach here. Observe the following.

First, as already noted in Chapters 9 and 10, it is a *very* rare occasion when the variance in the population is known; although in some exceptional circumstances one may have access to the population variance (e.g. from a previous study), in most instances we have to rely on the sample variance. Indeed, for all it is worth, the authors have never come across a situation where the population variance was known but the population mean was not. As the one-sample z-test is generally used when we know the population variance, its range of application is limited, particularly with small samples (where the assumption of a normal population must also hold).

With large samples, we can go ahead and use the z-test as an approximation, irrespective of the nature of the population distribution (we unashamedly did so in Chapter 10). However, recall from Chapter 9 that, as the sample size increases, the t-distribution (on which the t-test is based) follows more and more closely the normal distribution (on which the z-test is based). So, assuming that you have a reasonably-sized sample ($n>30$), the one-sample t-test and the one-sample z-test will give very similar results from a practical point of view.

Finally, in cases where the sample size is small *and* the population is decidedly non-normal, it is safest to go for a non-parametric test (such as the one-sample sign test discussed earlier) rather than risk being lynched by statistical hard-liners for using either the t-test or the z-test.

Bearing the above observations in mind, you should (a) use a one-sample *t*-test whenever you have a large sample, irrespective of the form of the distribution, (b) still use a *t*-test if you have a small sample but a normally distributed population, and (c) opt for a non-parametric test in all other instances. Phew!

A 'fringe benefit' of the *t*-test output (see Table 11.11), is that it contains all the information needed (i.e. sample mean, standard error, etc.) to calculate a confidence interval for the population mean (see Chapter 9). This is yet another opportunity to impress your client, boss, teacher, neighbour, spiritual advisor, etc. by casually demonstrating how to kill two birds with one stone (i.e. use the *t*-test output to do *both* hypothesis-testing *and* interval estimation).

Testing for Variability

Complementary to location tests are tests that allow us to examine hypotheses relating to the degree of *variability* in the population. Here the emphasis is on the extent of fluctuation or dispersion in the values of the variable of interest rather than on 'typical' or 'average' values. In this context, you should recall from Chapter 8 that the notion of variability is more meaningful when metric (i.e. interval or ratio) data are involved and that the most useful measure of variability is the standard deviation. Accordingly, we shall limit ourselves here to an illustration of a hypothesis test concerning a population standard deviation.

Let as assume that, unless you test the hypothesis that the standard deviation of the number of horror videos rented annually does not exceed 90, you will be unable to sleep at night. Thus, your null hypothesis is that $\sigma \leq 90$ and you want to use your survey data to test it against the alternative hypothesis that $\sigma > 90$. Now, here's the bad news: *SPSS/PC+* won't do it for you (and neither will most other statistical packages). Why not?, you may ask with righteous indignation.

There are two reasons for this most unsavoury situation. The first is that one cannot test a standard deviation directly; one must test the variance instead (don't ask us why, it's the statisticians who are to blame – again!). The second reason is that *SPSS/PC+* does not have a facility for doing a **variance test** (a shortcoming of many standard statistical packages). However, the good news is that a variance test is dead simple to do yourself. Here's how.

First, translate your hypothesis from standard deviation terms into variance terms. This involves simply squaring your hypothesized value, as the standard deviation is defined as the square root of the variance (see Chapter 8). Thus, in our illustrious example, the null hypothesis becomes $\sigma^2 \leq 8100$ and the alternative hypothesis $\sigma^2 > 8100$. Next, calculate the sample variance s^2 (yes, you can get this from any package, including *SPSS/PC+*). Finally, divide the sample variance, s^2, by the hypothesized population variance, σ^2, and multiply the ratio by the sample size n minus 1; in other words, calculate the following quantity:

variance test

$$(n-1)\frac{s^2}{\sigma^2}$$

The beauty of the above is that it follows a chi-square distribution with $(n-1)$ degrees of freedom and can thus be used as a test statistic; once you have calculated it, you can refer to tables of the chi-square distribution (you can find these in any statistics text, such as those suggested in the Further Reading section) and check out its significance. In our example, if we go back to Table 11.11 we see that $n = 110$ and $s^2 = (83.943)^2 = 7046.427$. Plugging these values in the above formula gives us

$$(110-1)\times(7046.427/8100) = 94.82$$

A chi-square value of 94.82 with 109 degrees of freedom is not significant (if you doubt this, check your statistical tables), which means that we cannot reject the hypothesis that the variance of horror videos rented annually is no more than 8100. Or, translating back to our original hypothesis, we can conclude that the standard deviation of horror videos rented does not exceed 90 videos.

Two points need to be mentioned in connection with the above procedure. First, the test makes the assumption that the underlying population is normal; violation of this assumption can lead to misleading results, as the test statistic will not be distributed as a chi-square (unfortunately, this test is not particularly robust). Second, when a two-sided hypothesis is involved, a complication arises in the calculation of the p-value because, unlike the normal and t-distributions, the chi-square distribution is not symmetrical (see also Chapter 8 on the notion of symmetry in distributions). This means that we cannot simply double the p-value if we want a two-tailed test. Instead, we have to follow the advice of statistical gurus, who suggest that 'for two-sided tests based on asymmetrical distributions, we may report the one-sided p-value accompanied by a statement indicating the direction of the observed departure from the null hypothesis'.[2]

WARNING 11.3

Make sure that the normality assumption is justified before you apply variance tests (see also Hint 11.1). Also remember that with asymmetrical distributions (such as chi-square), doubling of one-tailed p-values is *not* legitimate.

Testing for Proportions

Sometimes we want to test hypotheses concerning a variable that only takes two values (i.e. is dichotomous). An obvious example of such a variable is sex, with male and female being the two categories. Variables of this nature are naturally discussed in terms of *proportions* (or percentages), and if we know that the proportion of cases that fall into one category is P, then we also know that the proportion in the other category is $1-P$. For example, if the proportion of male students at the *Frankenstein Institute of Technology* in Düsseldorf is 0.78, then obviously the proportion of female students *must* be 0.22.

The kinds of hypotheses that can be tested when proportions are involved are directly analogous to those associated with location tests (see Table 11.7 earlier) and are summarized in Table 11.13. Probably the most widely tested hypothesis is when the population proportion, π, is set to 0.5, which is the same thing as saying that the proportions in both categories of the variable are equal in the population.

Table 11.13 Hypotheses for one-sample proportion tests

Null hypothesis	Alternative hypothesis	Type of test
$\pi = \pi_0$	$\pi \neq \pi_0$	2-tailed
$\pi \leq \pi_0$	$\pi > \pi_0$	1-tailed
$\pi \geq \pi_0$	$\pi < \pi_0$	1-tailed

π = population proportion, π_0 = hypothesized value.

binomial test

The statistical procedure you should use to test a hypothesis concerning a population proportion is known as the **binomial test**. The test has two versions, depending on the size of the sample, n, and the (hypothesized) population proportions π and $1-\pi$. When $n\pi$ or $n(1-\pi)$ is smaller than 5, the test uses the binomial distribution and provides an exact probability value ('small sample' case). When $n\pi$ and $n(1-\pi)$ are both greater than 5, the test uses the normal approximation to the binomial (see Chapter 9) to calculate a probability value ('large sample' case). Incidentally, the latter version of the binomial test is often referred to in standard statistics texts as

z-test for a proportion

the **z-test for a proportion** (see Figure 11.1 and Further Reading section). Moreover, if only to confuse matters even more, some texts discuss the 'large sample' case under parametric tests and the 'small sample' case under non-parametric tests. While we could all do without such complications, thankfully, most computer packages automatically determine from the input specifications which version of the test is appropriate.

To illustrate the application of the binomial test, let us test the hypothesis that the proportion of men who have a life subscription to *Horror Video Weekly* is 0.70. Thus, in null hypothesis terms we set $\pi = 0.70$, the alternative hypothesis being that $\pi \neq 0.70$. Moreover, let us assume that we have just conducted two separate telephone surveys using random samples of subscribers to this quality magazine; the first sample consists of 24 subscribers based in Yeovil and the second of 86 subscribers in Penzance. Tables 11.14 and 11.15 show the results of testing this hypothesis (*SPSS/PC+* output).

Table 11.14 An example of the binomial test: small sample

SEX

Cases

20	= 1		Test Prop.	= 0.7000
4	= 0		Obs. Prop.	= 0.8333
—			Exact Binomial	
24	Total		2-tailed p	= 0.2222

Table 11.15 An example of the binomial test: large sample

SEX

Cases

70	= 1		Test Prop.	= 0.7000
16	= 0		Obs. Prop.	= 0.8140
—			z Approximation	
86	Total		2-tailed p	= 0.0286

The output of the test is identical for the small and the large samples, the only difference being the distribution used to calculate the *p*-values. SEX is coded as a dichotomous variable, with 1 = male and 0 = female (of course we could have coded this variable using *any* two different numbers; if you are not sure why, go back to Chapter 3). The output shows both the *hypothesized* proportion in the population (referred to as 'Test Prop.') and the proportion *observed* in the sample (referred to as 'Obs. Prop.').

According to the findings for the Yeovil sample (Table 11.14), we cannot reject the null hypothesis that $\pi = 0.70$ as the test is non-significant ($p > 0.10$). However, the test returns a significant result (at $p < 0.05$) when applied to the Penzance sample (Table 11.15), indicating that the proportion of male subscribers to *Horror Video Weekly* is not 0.70. In fact, judging from the observed (i.e. sample) proportion, the relevant population proportion is likely to be greater than 0.70 (whether the conflicting results for the Yeovil and Penzance samples reflect differences in location-specific factors, differences in statistical power as a result of grossly unequal sample sizes, or both, is something that need not concern us here). Table 11.16 summarizes the steps involved in using the binomial test.

1. Specification of the proportion value and formulation of the desired null hypothesis (see Table 11.13).
2. Comparison of the hypothesized proportion with the observed proportion in the sample.
3. Examination of binomial exact probability (small samples) or normal approximation (large samples) and rejection (or non-rejection) of appropriate null hypothesis.

Table 11.16   Applying the binomial test

At this stage, you may be getting the strange feeling that the binomial test is rather similar to the one-sample chi-square test for testing goodness-of-fit (albeit for dichotomous variables). This is very much so and, provided that the assumptions underlying the one-sample chi-square test are met (see earlier discussion), one could translate a hypothesis regarding proportions into expected frequencies and perform a one-sample chi-square test. However, given the somewhat 'special' nature of dichotomous variables in terms of level of measurement (see Chapter 3), as well as their natural interpretation as proportions, the binomial test is more appealing. Its appeal is further enhanced by noting that both one- and two-tailed *p*-values can readily be obtained with the binomial test, whereas the asymmetric nature of the chi-square distribution makes matters a bit more complicated in this respect (see also Warning 11.3 above).

HINT 11.2

For testing hypotheses on dichotomous variables, the binomial test is preferable to the one-sample chi-square test.

Testing for Randomness

If you have made it this far, you will be pleased to know that there is only one more test to consider before you can put behind you this terrifying experience of single-variable, single-sample hypothesis tests (of course, this is nothing compared to the horrors awaiting you in Chapters 12 and 13!). The test in question is the **Wald–Wolfowitz runs test** for randomness (and we treat with contempt any suggestion that its name is indicative of the eating habits of its developers).

Wald–Wolfowitz runs test

Now, why would anybody in his/her right mind be interested in formulating and testing hypotheses concerning randomness? Goodness-of-fit, okay. Location and variability, sure, it makes sense. Proportions, well, all right. But *randomness?*

Recall from Chapters 2 and 6 that an important requirement in statistical inference (whether in the form of estimation or hypothesis testing) is *random* sampling. Thus it is quite handy to have a method for actually testing the hypothesis that our sample is indeed a random sample; this is precisely what the runs test allows us to do. Given a set of observations, the runs test examines whether the value of each observation influences the values taken by later observations; if the observations are independent, the sequence is considered random. More specifically, the test looks at the **runs** number of **runs** present in the sample, a run being defined as a sequence of like observations; too few or too many runs suggest dependence between observations (i.e. lack of randomness).

Consider the following three sequences reflecting the sex of 10 respondents queuing at a cinema ticket counter (for the premiere of *Nightmare on Downing Street III*):

MMMMM FFFFF
M F M F M F M F M F
FFF MM F MM F M

In the first two sequences (containing two and ten runs respectively), we suspect a lack of randomness, as there appears to be a pattern running through the observations (i.e. 'bunching' and 'alternating' of the sexes, respectively). In contrast, the third sequence (containing six runs) appears to be well mixed and gives little cause for suspecting a lack of randomness.

Now that you have the general idea, let us apply this test to check out whether your sample of 24 Yeovil subscribers to *Horror Video Weekly* (see Table 11.14 earlier) is indeed a random one with regard to respondent sex. Table 11.17 shows the results (yawn, yawn, it's *SPSS/PC+* output again).

Table 11.17   An example of the runs test

SEX

Runs:	9		Test value	= 1
Cases:	4	Lt 1		
	20	Ge 1	z	= 0.6502
	24	Total	2-tailed p	= 0.5155

Lt = less than, Ge = greater than or equal to.

The 'Test Value' in the output indicates the way in which the runs are defined; in this case this has been set equal to 1, as we have coded SEX 1 = male, 0 = female (and we are interested in seeing whether the sequence of male and female respondents in our sample of 24 individuals is indeed random). On this basis, there was a total of nine runs and, according to the test statistic, this number was neither too low nor too high to suggest non-randomness (the z-value is non-significant). Thus we cannot reject the hypothesis that, as far as respondent sex is concerned, the sample is random (which is just as well, otherwise we would have trouble justifying the use of this sample in our previous analyses).

When the variable involved is dichotomous (as in the above example) the runs are straightforward to define. What happens if the variable is continuous (or discrete but with many values)? No problem – the runs test can cope with such data just as well. All one has to do is specify a 'cutting point' for dichotomizing the variable being tested (usually, the mean, median or mode). This cutting point then becomes the 'Test Value', i.e. values less than the cutting point form one category and values greater than or equal to the cutting point form the other category. Then the runs are calculated exactly as before. Table 11.18 shows an application of the runs test to a continuous variable with the mean being used as the 'Test value'; the variable concerned is the age of the 24 Yeovil subscribers to *Horror Video Weekly*.

AGE				
Runs:	14		Test value	= 41.04 (Mean)
Cases:	13	Lt Mean		
	11	Ge Mean	z	= 0.5196
	24	Total	2-tailed p	= 0.6033

Lt = less than, Ge = greater than or equal to.

Table 11.18 Another example of the runs test

You should note with great delight that the usefulness of the runs test is not restricted to testing the null hypothesis that a given sample is random (although this is its most common application). The test can be applied to any sequence, no matter how the sequence was generated. For example, you could apply it to test for randomness of the presence or absence of snow over a given period, the wins and losses of the local rugby team, the correct and false answers to a multiple choice test, and so on. In fact, whenever you see a sequence, look at it as a not-to-be-missed opportunity for applying the runs test.

Summary

In this chapter we took the first concrete steps into hypothesis-testing by examining various hypotheses relating to a single variable and a single sample. We started by looking at hypotheses concerned with goodness-of-fit and had a jolly good time fooling around with the one-sample chi-square test and the Kolmogorov–Smirnov test. We then moved on to testing hypotheses regarding location and had a ball with the one-sample sign test and the one-sample t-test. Hypotheses concerning variability were next in line and these were swiftly dealt with (*without* a computer, may we add) via the variance test. Dichotomous variables were tackled subsequently and the binomial test was used to test hypotheses concerning proportions. The icing on the cake was none other than the (somewhat unfortunately named) runs test which is *the* test to use where randomness is the issue. We are now ready to enter more complex territory and have some fun making comparisons between groups and between measurements.

Questions and Problems

1. Explain in non-mathematical terms the underlying hypotheses for the following tests:

   (a) goodness-of-fit tests
   (b) tests for location
   (c) tests for variability
   (d) tests for proportions
   (e) tests for randomness

2. What type of measurement does a variable require to apply the one-sample chi-square test?

3. What is the null hypothesis under the one-sample chi-square test?

4. What is meant by the terms 'theoretical frequencies' and 'expected frequencies'?

5. Does the one-sample chi-square test require that all the expected frequencies are identical?

6. Which test is commonly applied to find out whether observed values have come from a normally distributed population?

7. Explain when you would use (a) a one-sample sign test, and (b) a one-sample *t*-test.

8. You have taken a random sample of 70 Buenos Aires taxi drivers and found that 90% admit to exceeding the speed limit by 60 miles per hour on a regular basis. Which test would you use to check whether this result is likely to apply to all taxi drivers in Buenos Aires?

9. In which situations (other than to impress friends at cocktail parties) would you use the Wald–Wolfowitz test?

10. Have you been to any good horror movies lately?

Notes

1. Guildford, JP (1965) *Fundamental Statistics in Psychology and Education*, p. 262. London: McGraw-Hill.
2. Daniel, WW & Terrell, JC (1995) *Business Statistics for Management and Economics*, 7th edn, p. 368. Boston: Houghton Mifflin.

Further Reading

Hinkle, DE, Wiersman, W & Jurs, SG (1988) *Applied Statistics for the Behavioral Sciences*. Boston: Houghton Mifflin. An all-round statistics text which is nicely structured and not too technical. An added bonus is that it includes computer examples and exercises for *SPSS* and *SAS* (although for mainframe rather than PC versions).

Siegel, S & Castellan, NJ (1988) *Nonparametric Statistics for the Behavioral Sciences*, 2nd edn. New York: McGraw-Hill. This the second edition to the 'bible' on non-parametric statistics, first published in 1956 (ah, those were the days!). Essential reading.

Tulle, DS & Hawkins, DI (1993) *Marketing Research: Measurement and Method*, 6th edn. New York: Macmillan. Chapters 19–20 contain a variety of statistical tests presented in an easy-to-follow fashion, including several that we will be discussing/referring to in the next two chapters.

12

Getting experienced: making *comparisons*

The Pleasure of Comparing

Now that you have mastered the fundamental ideas about data analysis and are able to squeeze various statistical tests out of your computer, it is time to apply your skills to something more pleasurable, namely comparisons. In fact, it is often only comparisons which provide meaning. Concepts like small and large, young and old, or ugly and beautiful, only make sense if used in a comparative sense (for example, a pot-bellied pig is uglier than a rhinoceros). Similarly, figures like the sales revenue of a waterbed manufacturer in Tombstone (Arizona), the proportion of Italian men using wallpaper strippers or the number of US pet dog owners who buy them special Christmas dinners, become more illuminating if compared to, for example, waterbed manufacturers in Fishguard (Wales), men using wallpaper strippers in Angola and the number of Korean pet dog owners who eat them for Christmas (we are still working on the last example!). In all these examples, *different groups* are compared on a particular characteristic.

Comparisons can also involve contrasting *multiple measures*; for example, comparing the average importance given to different criteria when buying a car on a 5-point rating scale (e.g. price, fuel consumption, number of windscreen wiper speeds, and so on). Here, a single group (i.e. sample of respondents) provides two or more measurements which are then contrasted. Often, such measurements are obtained at different points in time and are aimed at detecting change. A typical example of such a comparison involves data of the 'before and after' variety; for example, comparing the percentage of Scotsmen suffering from diarrhoea before and after eating 11 portions of haggis, or attitudes towards smoking before and after being exposed to a 'shock' television advert (depicting an elderly smoker spontaneously combusting as a result of lighting up!).

From the above it should be clear that comparisons can take two basic forms: either two (or more) different *groups* are compared on a given characteristic or two (or more) *measures* of the same group are compared to one another. This distinction is important because it has direct implications for the choice of statistical technique. As extensively discussed in Chapter 6, **independent measures** with **independent measures** (or 'independent samples' as they are confusingly referred to in the statistical literature), different units of analysis **related measures** are compared on the same variable. In contrast, with **related measures** (or 'related samples'), the same units are compared on different variables. As Figure 12.1 shows, different techniques are appropriate for making comparisons between independent measures as opposed to related measures; note that we have only listed the most important and widely used techniques and we refer you to the Further Reading section if what's on offer in Figure 12.1 is not enough for you.

In what follows, we shall first explain the delights of comparing independent measures (in our experience, the most common situation) and then, more briefly, introduce you to techniques for undertaking comparisons between related measures. Since you are now 'old hands' at hypothesis-testing and given the number of techniques that we must try to cover (see Figure 12.1), we shall go through the various tests in a bit less detail than we did in Chapter 11; this should make this chapter more palatable and also avoid unnecessary repetition of points already covered.

WARNING 12.1

Before making comparisons you should establish whether you are dealing with independent or related measures. Failure to do this is likely to result in incorrect choice of statistical technique.

Independent Measures: Comparing Groups

This is a very typical situation in data analysis which occurs whenever (a) one has data from more than one sample (e.g. from two separate surveys of Internet 'surfing' habits of Algerian and Czech dentists), or (b) one decides to split a single sample into two (or more) sub-samples on the basis of some characteristic (e.g. creating sub-samples of male and female Algerian dentists and subsequently comparing their Internet surfing habits). In both cases, there are two or more *groups* involved and the interest lies in identifying similarities and differences between them. Clearly, any observed differences are only of importance if they are likely to apply to the respective *populations*. If it is only sampling error which is responsible for the patterns observed, then any conclusion that the groups compared differ with respect to the characteristic of interest is unwarranted (see Chapter 10). Thus, in comparing groups, it is necessary not only to identify differences based on the sample data but also to test the statistical significance of such differences.

In hypothesis-testing terms, comparisons between groups can be conveniently expressed as follows:

- There is no difference between the two (or more) groups in terms of the characteristic of interest (null hypothesis).
- The groups differ with respect to the characteristic of interest (alternative hypothesis – exploratory).
- One group has more/less of the characteristic of interest than the other group(s) (alternative hypothesis – directional).

Note that you should always compare the *same* variable (i.e. characteristic) between the groups; that is do not attempt to compare the average age of Italians taking sleeping pills before going to bed with the average length of marriage among Germans who take sleeping pills. Instead, conduct two analyses, one comparing the average age and one comparing the average length of marriage between the groups.

Depending upon the nature of the characteristic which is the focus of comparison, hypotheses can be formulated relating to frequencies

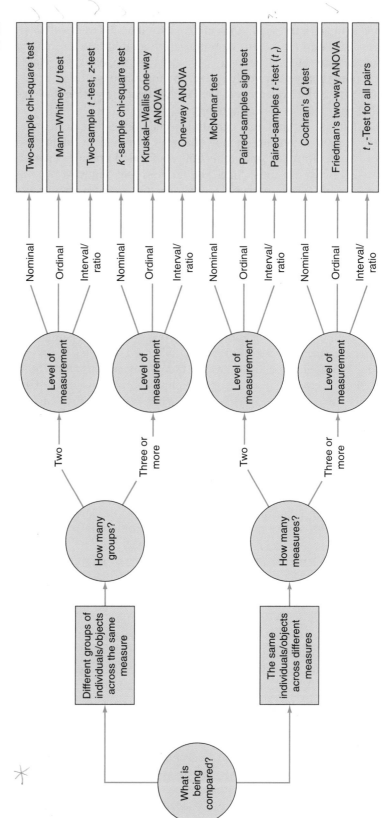

Figure 12.1 Statistical techniques for making comparisons.

(nominal data), rank orders (ordinal data), and mean levels (interval/ratio data) across the groups. Unfortunately, as is clear from Figure 12.1, the level of measurement of the characteristic of interest is not the only determinant of the appropriate statistical technique for carrying out the comparison. The *number* of groups to be compared also needs to be taken into consideration, since different techniques are used for two-group comparisons (e.g. comparing a sample of British households owning electric curling tongs with an equivalent sample from Cameroon) from those for *k*-group comparisons (e.g. comparing Swedish, Norwegian and Finnish owners of pet tarantulas). While you may be inclined to have a nervous breakdown at this stage as even more tests have to be learned, there is a bright side – really! All the tests for making *k*-sample comparisons (where *k*>2) are straightforward extensions of the equivalent tests for making two-sample comparisons. What is even better, the two-sample tests are themselves often based on the single-sample tests that you so enjoyed learning about in Chapter 11. So, please do not give up – make yourself a herbal tea or go get a massage and, then, read on.

Now that we have lulled you into a false sense of security, we can start looking into different statistical tests for making comparisons. In order to do this as painlessly as possible, we will first discuss the two-group case for each level of measurement and immediately follow it with the *k*-group case. To ease your suffering even further, we will set $k = 3$ for our examples, i.e. deal with three groups only (this results in no loss of generality in that you can use exactly the same techniques to compare any number of groups you like – assuming, of course, that the relevant test assumptions are satisfied). If during the following discussion you feel that you are getting lost or slowly drifting to sleep, a quick glance at Figure 12.1 will remind you where you are.

The Two-Sample Chi-Square Test

If you want to compare two groups on a variable which is measured on a nominal scale, this is the test you need. As its name implies, the **two-sample chi-square test** is the big brother (or big sister, if you prefer) of the one-sample chi-square test (see Chapter 11) and, like its baby sibling, is based on a comparison of observed versus expected frequencies.

two-sample chi-square test

The null hypothesis tested by the two-sample chi-square test is that no difference exists between the two groups with respect to the *relative* frequency with which group members fall into the various categories of the variable of interest. The reason for focusing on relative rather than absolute frequencies is that the two groups may have unequal sample sizes and, therefore, the calculation of expected frequencies needs to take this into account. If the observed frequencies depart significantly from the expected frequencies (i.e. cannot be dismissed due to sampling error), we conclude that the two groups differ along the variable of interest. If, on the other hand, we find that the discrepancies between observed and expected frequencies are small and non-significant, then we have no evidence of differences between the groups.

The best way to visualize all this is by means of a **contingency table**, that is the cross-tabulation you obtain from tabulating your group variable

contingency table

against the characteristic of interest. For example, if you want to compare Greeks and Nigerians on a variable called 'natural headache remedies' (indicating their typical approach to curing headaches), you could visualize the data as illustrated in Table 12.1 (yes, of course, this output has been produced by the *SPSS/PC+* program – would you expect us to do the work ourselves?).

Table 12.1   An example of the two-sample chi-square test

HEADACHE BY COUNTRY

		COUNTRY		
		Greece	Nigeria	
	Count Exp. Val.	1	2	Row total
HEADACHE Sleeping	1	584 581.5	100 102.5	684 46.8%
Drinking Ouzo	2	599 598.5	105 105.5	704 48.2%
Juju Music	3	59 62.1	14 10.9	73 5.0%
	Column total	1242 85.0%	219 15.0%	1461 100.0%

Chi-square	Value	D.F.	Significance
Pearson	1.08148	2	0.58232
Likelihood Ratio	1.01452	2	0.60214
Mantel–Haenszel test for linear association	0.48819	1	0.48474

Minimum expected frequency: 10.943
Number of missing observations: 12

From Table 12.1 you can see that the Greek sample consists of 1242 people, whereas the Nigerian sample has only 219. The column percentages indicate the relative composition of the sample (i.e. 85% Greeks and 15% Nigerians, respectively), whereas the row percentages indicate the overall popularity of each of the three headache remedies across both samples (with 'drinking ouzo' being the most popular and 'Juju music' least popular). Within each of the six cells of the table, two further pieces of information are given: firstly, the number (i.e. observed frequencies) of Greeks and Nigerians who are curing their headaches with one of the three remedies and, secondly, the number of individuals in each sample that would be expected to prefer each remedy if there was no difference in preferences between the two groups (i.e. the expected frequencies). The latter are denoted as 'Exp.Val.' and are simply calculated by multiplying each *row* percentage with the respective sample sizes; for example, the expected frequency of Greeks opting for 'Juju music' as a headache cure is 5% of 1242, i.e. 62.1 (see bottom-left cell of Table 12.1).

So, where does the chi-square test come in? It comes in the comparison of actual versus expected frequencies because, if the null hypothesis is true (see earlier), there should be no difference between the actual and expected frequencies in each cell of the table. In our example, the (Pearson) chi-square statistic comes to 1.081, with an observed probability of 0.582. Since the latter is not smaller than 10% ($p<0.10$), 5% ($p<0.05$) or 1% ($p<0.01$), which are the usual cut-off points for rejecting the null hypothesis (see Chapter 10), we would conclude that there are no statistically significant differences between the Greek and Nigerian populations in terms of their favourite headache remedies (simply ignore the other two statistics in Table 12.1 – the likelihood ratio and the Mantel–Haenszel test – *SPSS/PC+* produces them by default but they need not concern you here). Had we obtained a larger chi-square value and a lower probability (e.g. a chi-square of 7.378 with a probability of 0.025), we would have been able to reject the null hypothesis. Specifically, we would have argued that the differences between observed and expected frequencies are so large that they would occur only 2.5% of the time if, in the respective populations, the three headache remedies were equally popular. Altogether very easy and straightforward!

However, before you go out and celebrate at your local Senegalese restaurant that you understood when and how to use the two-sample chi-square test, there are a few caveats which you have to remember. The chi-square test may not be correct if the expected frequencies are less than 5. In fact, you should always ensure that no more than 20% of the cells in your table have expected values of less than 5 and none of the expected values is less than 1 (see also Chapter 11 on similar limitations of the one-sample chi-square test). Try to combine adjacent categories (i.e. reduce the size of the table) if you face such difficulties making sure, however, that such combined categories are still meaningful; for example, two separate categories for whiskey and gin could be combined into a new category called 'spirits', but it would be difficult to extract meaning from a combination of ouzo and juju music. As you can see from Table 12.1, the computer output indicates the 'Minimum Expected Frequency' so you get an early warning about potential problems with small expected frequencies (in our example, this comes to 10.943, so no problems).

With a 2×2 table (i.e. when two groups are compared on a dichotomous variable), it is often recommended (for reasons far too esoteric to explain here) that **Yates' correction for continuity** is applied to obtain a modified chi-square statistic (the latter is somewhat lower than the unmodified chi-square and, thus, results in a more conservative test). Fortunately, statistical packages such as *SPSS/PC+* routinely apply this correction if needed, so you do not have to worry about it (see later example in Table 12.2). This is just as well, since 'statisticians aren't in agreement as to whether this correction is really necessary'![1]

While on the thrilling subject of 2×2 tables, it takes no great imagination to realize that if you have a small sample (say 20 cases or fewer), the requirements associated with minimum expected frequencies are *bound* to be violated. As it is not possible to combine categories (you would be left without a table), what can you do? Three avenues are open to you, namely (1) go into the first pub and drown your sorrows, (2) burst into tears and

WARNING 12.2

In applying the two-sample chi-square test, ensure that no more than 20% of the cells have expected frequencies of less than 5, and no cell has an expected frequency of less than 1.

HINT 12.1

If more than 20% of the cells have expected frequencies of less than 5 and/or at least one cell has an expected frequency of 1 or less, try to combine categories so as to reduce the number of cells in the contingency table.

Yates' correction for continuity

hope that your supervisor/boss/customer takes pity on you, or (3) use the so-called **Fisher's exact test**. This test is a godsend when you have a small sample as it evaluates exactly the same null hypothesis as the chi-square test. Fisher's exact test is based on the **hypergeometric distribution**, which is yet another discrete probability distribution dreamed up by statisticians (yes, they have done a good job of coming up with enough distributions to drive us all around the bend!). The output of the test is simply an exact probability value which can be compared to a pre-set significance level (e.g. 5%) to determine whether significant differences exist between the groups. Being incredibly clever, computer packages tend to perform this test automatically when required; the *SPSS/PC+* package does so whenever the sample size in the 2×2 table is 20 or less.

Table 12.2 shows an example of the Fisher exact test applied to a cross-tabulation of sex (SEX) by marrying intentions (REMARRY); the data relates to 18 Hungarian divorcees who were asked to indicate whether they intended to marry again within the next six months (note that one respondent rudely refused to answer the question and was thus promptly relegated to a 'missing observation').

Table 12.2 An example of Fisher's exact test

REMARRY BY SEX

		SEX		
		Male	Female	
REMARRY	Count Exp. Val.	0.00	1.00	Row total
No	0.00	6 4.7	2 3.3	8 47.1%
Yes	1.00	4 5.3	5 3.7	9 52.9%
	Column total	10 58.8%	7 41.2%	17 100.0%

Chi-square	Value	D.F.	Significance
Pearson	1.63254	1	0.20135
Continuity correction	0.61473	1	0.43301
Likelihood Ratio	1.67214	1	0.19597
Mantel–Haenszel test for linear association	1.53651	1	0.21514
Fisher's exact test:			
One-tail			0.21781
Two-tail			0.33484

Minimum expected frequency:	3.294	
Cells with expected frequency:	<5	3 of 4 (75.0%)
Number of missing observations:	1	

Although the value of the chi-square test is still provided in the output, the fact that three out of four cells have expected frequencies less than 5 indicates that it should not be relied upon (either in its raw form or after adjusting for continuity). Instead, we take a look at the probability associated with Fisher's exact test; as neither the one-tailed nor the two-tailed probabilities are remotely close to reaching significance at conventional levels, we can conclude that male and female divorcees do not differ with respect to their remarrying intentions.

A useful feature of chi-square analysis of 2×2 tables is that it enables the testing of hypotheses concerning differences in *proportions* between two populations (see Table 10.1 in Chapter 10 for a summary of possible hypotheses). Given that there are two groups involved and a dichotomous variable of interest, the calculation of expected frequencies reflects a null hypothesis of equal proportions between the two groups (see earlier points on the computation of expected frequencies). Although there is another technique that one can use to test for differences between two population proportions (namely a **z-test for differences in proportions** based on the normal distribution – see Further Reading sources for details), 'when there are two populations involved and the characteristic of interest has two categories, the chi-square test of homogeneity is an alternative and equivalent way of testing the null hypothesis that two population proportions are equal'.[2] Indeed, the chi-square test has the advantage that it can be applied to small samples, whereas the z-test for differences in proportions should only be applied if *each* group has 30 cases or more. On the other hand, the z-test is more flexible as the symmetrical nature of the normal distribution on which it is based allows both one-tailed and two-tailed tests to be carried out, whereas the chi-square test does not (see Warning 11.3 in Chapter 11). So, in the end, it's all swings and roundabouts.

z-test for differences in proportions

The *k*-Sample Chi-Square Test ✓

We unreservedly apologize for the above heading: '*k*-sample' sounds positively dreadful and likely to leave you with a bitter aftertaste in your mouth! However, statistics and marketing research books frequently use this expression when they mean comparisons between three or more groups and, therefore, we thought that you should know what they are on about.

The **k-sample chi-square test** is nothing but an extension of the two-sample chi-square test when more than two groups need to be compared on a nominal variable. Thus if, in a flash of inspiration, we extended our analysis of natural headache remedies by including a sample from Sri Lanka – in addition to our Greek and Nigerian samples – our null hypothesis would be that there are no differences between the three groups with respect to the relative frequency with which group members prefer the various remedies. The approach for setting up the contingency table and for calculating expected frequencies would be exactly the same as for the two-group case and so would the computer output. Specifically, a chi-square statistic would be provided together with a probability of occurrence under the assumption that the null hypothesis was true; provided that this probability was smaller than our significance level, we would

k-sample chi-square test

reject the null hypothesis and conclude that the three populations differ in terms of preferred headache remedies.

The two-sample and *k*-sample chi-square tests are often referred to as **tests of homogeneity**, since they test whether several groups are homogeneous with regard to the characteristic of interest. Alternatively, they can be seen as **tests of independence** from the perspective of testing whether the two (nominal) variables forming the contingency table are related or not. While the two interpretations are not strictly synonymous (for reasons better not delved into), from a practical viewpoint they can be treated as such.

tests of homogeneity

tests of independence

As with the two-sample chi-square test, no fewer than 20% of the cells should have an expected frequency of less than 5, and no cell should have an expected frequency smaller than 1 (see Warning 12.2 and Hint 12.1 above). With a fixed sample size, the bigger the table (i.e. the more groups that are compared and/or the greater the number of categories of the variable of interest), the greater the chances that you will run into problems. Bear this in mind when deciding on your sample size (see also Hint 2.1 in Chapter 2).

HINT 12.2

Look at the largest cross-tabulation (in terms of rows × columns) that you want to analyse *before* you collect data and adjust your sample size accordingly (or lower your ambitions!).

Now that you feel at home with all possible versions of the chi-square test (single-sample, two-sample, *k*-sample), here's some food for thought: the value of the chi-square statistic is dependent upon the sample size. In other words, without changing the pattern in the data, you can get a significant result simply by increasing your sample. To illustrate this, if we multiply all the entries in Table 12.1 by 20, the chi-square value would also become 20 times greater and highly significant (a chi-square value of 21.63 with 2 degrees of freedom is associated with a *p*-value less than 0.0001!). This happens because statistical power increases dramatically as a result of the increase in sample size (see Chapter 10) and the test goes mad and declares as significant even minuscule differences between the two groups. Indeed, with large sample sizes, even very small differences between two groups may be statistically significant. It is therefore advisable to check the percentages in the table to determine whether the statistically significant results are of any practical relevance.

WARNING 12.3

When using the chi-square test in group comparisons, bear in mind its dependence on sample size and go beyond an investigation of statistical significance; look at the *magnitude* of any revealed 'differences'.

Finally, let us enlighten you with regard to the 'DF' bit that is always part and parcel of any chi-square test output. This stands for 'degrees of freedom' and is always equal to $(r-1)(c-1)$, where r = number of rows and c = number of columns in the contingency table. Thus in our example in Table 12.1, DF = 2 given that there are three rows and two columns (since $(3-1)\times(2-1) = 2$), whereas in Table 12.2, DF = 1 as there are only two rows and two columns. The case of a single-sample chi-square test (discussed in Chapter 11) can be considered as a contingency table having only one row (or one column); the associated degrees of freedom are $r-1$ (or $c-1$). Indeed, if you go back to Tables 11.1 and 11.2 you will see that there are three degrees of freedom, i.e. one less than the number of categories in the variable of interest.

The Mann–Whitney *U* Test

Mann-Whitney *U* test

The **Mann–Whitney *U* test** (also known as the 'Wilcoxon rank sum *W* test') is very useful when you have two groups to compare on a variable which is measured at ordinal level. The test focuses on differences in central location and makes the assumption that any differences in the distributions of the two populations are due only to differences in locations (rather than, say, variability).

The null hypothesis tested by the Mann–Whitney U test is that there is no difference between the two groups in terms of location, focusing on the median as a measure of central tendency. Note that, in some statistics texts, the null hypothesis is simply stated as involving identical distributions for the two populations; this amounts to the same thing because, as noted above, the test assumes that, if they differ at all, the distributions of the populations differ only with respect to location. You should also bear in mind that, in the case of symmetrical distributions and given interval data, the test can also be used to draw conclusions about means (since the mean and median coincide in symmetrical distributions – see Chapter 8). Thus the Mann–Whitney U test is a useful alternative to a parametric location test (such as the two-sample t-test discussed below) when assumptions about normality are violated and/or the sample sizes are small.

Consider the data in Table 12.3, indicating the quality rankings assigned by a sample of 47 Catholic padres and a sample of 56 Jewish rabbis to the *Happy Nunnery* (a disco specializing in entertainment for clergy); respondents were asked to rank four well-known discos, where 1 = best and 4 = worst.

Table 12.3 Quality rankings for the Happy Nunnery

HAPYRANK BY RELGROUP

HAPYRANK	Count	Padres (1)	Rabbis (2)	Row total
1		27	4	31 / 30.1%
2		10	5	15 / 14.6%
3		5	16	21 / 20.4%
4		5	31	36 / 35.0%
Column total		47 / 45.6%	56 / 54.4%	103 / 100.0%

We can see that 31 respondents placed the *Happy Nunnery* in the number one position, 15 in the number two position, etc. If we want to find out whether the assigned quality rankings differ between Catholic padres and Jewish rabbis, the Mann–Whitney U test provides the answer; the relevant computer output (generously provided by *SPSS/PC+*) is shown in Table 12.4.

Table 12.4   *An example of the Mann–Whitney U test*

HAPYRANK
by RELIGION

Mean rank	Cases	
32.65	47	RELGROUP = 1 Padres
68.24	56	RELGROUP = 2 Rabbis
	103	Total

Corrected for ties

U	W	z	2-tailed p
406.5	1534.5	−6.2832	0.0000

The first piece of information to note from the output is the 'Mean rank' for each group (i.e. the sum of the ranks divided by the number of cases). If there are no differences in the populations, we would expect similar ranks in the two groups; if either group has more of its share of either large or small ranks, then it would be unlikely that the respective populations would be the same. The W statistic is simply the sum of ranks for the group with the smaller number of cases (indeed, if you divide W by the sample size of the smaller group you will come up with its mean rank; go ahead, try it yourself). The U statistic shows how often a ranking of the group with the *largest* number of cases is smaller than a value in the other group. Note that the significance levels of the W and U statistics are the same and, provided the *total* sample size is greater than 30, they are obtained through a transformation to a z-value (i.e. a standard normal deviate); the latter is 'Corrected for Ties', that is it includes an adjustment for cases with tied (i.e. equal) ranks. If the total sample size is less than 30, an exact significance level for W and U is also displayed (as the z-transformation may not be accurate with small samples).

The results in Table 12.4 are highly significant ($p<0.0001$) and we can therefore reject the null hypothesis that the quality rankings of the two groups are the same. Moreover, remembering that the mean ranks represent the sum of the ranks divided by the number of cases, we can see that our 47 Catholic padres are ranking the *Happy Nunnery* higher than the 56 Jewish rabbis (recall that 1 represents the highest rank). Since some of you might not believe us (and who could blame you for doubting the authenticity of our examples!), look again at the cross-tabulation in Table 12.3 and you will see with surprising clarity that our padres 'dig' the *Happy Nunnery* more than our rabbis.

The Kruskal–Wallis (K-W) One-Way Analysis of Variance (ANOVA)

Kruskal–Wallis one way analysis of variance

If you wish to become even more adventurous and decide to compare an ordinal variable (e.g. preference rankings of films, advertisements, hamburger joints or, indeed, discos) across three or more independent groups, the **Kruskal–Wallis one-way analysis of variance** is the test for you. In this context, you should note that the mere mention of the name of this test will, in the right circles, immediately qualify you as an intellectual (is this the way you spell intellectual?). In contrast, the authors experienced remarkably little success at parties with chat-up lines such as 'Can I possibly interest you in a Kruskal–Wallis one-way analysis of variance?'

In any event, the K-W one-way ANOVA tests the same null hypothesis as the Mann–Whitney U test but across k rather than two groups (i.e. all groups have the same distribution in the population and any differences reflect differences in location only). In a manner similar to the use of the Mann–Whitney U test as a non-parametric alternative to the t-test for differences between two means, K-W one-way ANOVA can be considered as a viable alternative to the parametric one-way ANOVA procedure (to be discussed later in this chapter), when the assumptions of the latter are violated.

To illustrate the use of the K-W one-way ANOVA, we have extended our survey of padres and rabbis by asking a sample of 42 Buddhist monks (doing missionary work in Las Vegas), what they think about the *Happy Nunnery* discotheque; the new data are shown in Table 12.5, while Table 12.6 displays the results of the associated K-W one-way ANOVA run (courtesy of *SPSS/PC+*).

HAPYRANK BY RELGROUP

		RELIGION			
		Padres	Rabbis	Monks	
	Count	1	2	3	Row total
HAPYRANK	1	27	4	9	40 27.6%
	2	10	5	18	33 22.8%
	3	5	16	7	28 19.3%
	4	5	31	8	44 30.3%
	Column total	47 32.4%	56 38.6%	42 29.0%	145 100.0%

Table 12.5 Quality rankings for the Happy Nunnery – three groups

The first thing that attentive readers (yes, they *do* exist) will have noticed from Table 12.6 is that the 'Mean Ranks' given for padres and rabbis differ from those given in the Mann–Whitney U results in Table 12.4, although the quality rankings of these two groups have not been changed. There is a technical reason for this, namely that the test computations rest on the combined ranks of *all* groups. Since we added the 42 observations of the Buddhist monks, this has affected the mean ranks of the padres and rabbis. Of course, their actual assessment of the Happy Nunnery disco has

Table 12.6 An example of Kruskal–Wallis one-way ANOVA

HAPYRANK				
by RELGROUP				
Mean rank	Cases			
46.35	47	RELGROUP = 1 Padres		
99.92	56	RELGROUP = 2 Rabbis		
66.93	42	RELGROUP = 3 Monks		
	—			
	145	Total		

			Corrected for ties	
Cases	Chi-square	Significance	Chi-square	Significance
145	42.8006	0.0000	45.9174	0.0000

not changed, as can be verified by comparing the cross-tabulations *with* (Table 12.5) and *without* (Table 12.3) the Buddhist monk sample added.

As can be seen from the test statistic, the results are again highly significant ($p<0.0001$) and we can, therefore, reject the null hypothesis that there is no difference in the ranking of the *Happy Nunnery* between the three groups; indeed, looking at the mean ranks, padres think the most of the *Happy Nunnery*, while rabbis the least; the rankings of monks are in between.

The test statistic associated with K-W one-way ANOVA is based on an approximation of the chi-square distribution with $k-1$ degrees of freedom, where k is the number of groups compared; an adjustment, taking into account tied ranks, is also provided (this is denoted as 'Corrected for Ties' and is the figure one should generally use). If $k = 3$ and if any of the samples has five cases or fewer (which is a disgracefully low sample size), instead of the chi-square statistic, a so-called H statistic is computed together with an exact level of significance; the interpretation of the results remains, of course, the same.

The Two-Sample *t*-Test

This is a straightforward extension of the one-sample *t*-test for a mean (discussed in Chapter 11) and should be used when one wants to compare two groups on a variable measured at interval (or ratio) level. Thus, whenever you are overcome with an urgent desire to compare the means of two **two-sample *t*-test** groups, don't think twice: go for the **two-sample *t*-test**.

The null hypothesis tested by the two-sample *t*-test is that the two population means are equal (i.e. $H_0: \mu_1 = \mu_2$), the alternative hypothesis being that the means are not equal ($H_1: \mu_1 \neq \mu_2$). Of course, directional hypotheses can also be tested if one mean is *a priori* expected to be higher or lower than the other; the relevant hypotheses are then $H_0: \mu_1 \leq \mu_2$ versus $H_1: \mu_1 > \mu_2$ and $H_0: \mu_1 \geq \mu_2$ versus $H_1: \mu_1 < \mu_2$, respectively. Note that these hypotheses are analogous to those relating to one-sample location tests as summarized in Table 11.7 in Chapter 11.

Let us assume that you live in lovely Schweinebratenhausen (beautifully located in the Tyrolean Alps) and you want to compare the mean age of the people living in your town with the mean age of residents of Underwood, North Dakota (that's in the United States for those of you who haven't had the chance to visit yet). The total lack of previous research on this important topic leads you to postulate an exploratory hypothesis, namely that there is no difference in the average ages of the two groups. Having drawn random samples of residents in the two towns, you then confiscate their birth certificates and determine their ages. Subsequently, you diligently input your data into your computer and ask *SPSS/PC+* to perform a *t*-test for two independent samples; the output you get is shown in Table 12.7.

Independent samples of TOWN

Group 1: TOWN EQ 0.00   Group 2: TOWN EQ 1.00

t-test for: AGE age (in years)

	Number of cases	Mean	Standard deviation	Standard error
Group 1	50	42.3000	10.164	1.437
Group 2	51	39.2745	8.268	1.158

		Pooled variance estimate			Separate variance estimate		
F value	2-tail *p*	*t*-value	Degrees of freedom	2-tail *p*	*t*-value	Degrees of freedom	2-tail *p*
1.51	0.149	1.64	99	0.104	1.64	94.30	0.105

Table 12.7   An example of the two-sample *t*-test

The top portion of the output shows the means, standard deviations and standard errors for the two groups; the group variable is denoted as TOWN and coded as 0 for Schweinebratenhausen and 1 for Underwood. The lower portion of the output shows the results of the actual *t*-test based on (a) a 'Pooled variance estimate' and (b) a 'Separate variance estimate'. In its purest form, the two-sample *t*-test is based on the assumption that the variances of the two groups *in the population* are the same; if this assumption holds, then we look at the *t*-value, number of degrees of freedom (which equal $n-2$, where n = total sample size) and two-tailed probability under the 'Pooled variance estimate' heading. If the 'equal variances' assumption does not hold, then we look at the information under the 'Separate variance estimate' heading (the latter includes an adjustment in the degrees of freedom to take account of the inequality of variances).

But how do we *know* whether the two groups are likely to have the same population variances or not? In anticipation of this, good old *SPSS/PC+* provides us with a test, namely the so-called **F-test for equality of variances**; this involves forming the ratio of the two sample variances and is based on the **F-distribution** (yet another probability distribution with known properties – see Chapter 8). If the *F*-value is close to unity, the

F-test for equality of variances

F-distribution

sample variances are similar; the larger the F-value, the more dissimilar are the sample variances. The probability beside the F-value indicates the likelihood of seeing a ratio at least as large as the one observed in the sample data (here $F = 1.51$) if, in reality, the variances are equal in the population. If the F-value is significant, we must use the t-test results provided under the 'Separate variance estimate'; otherwise, we can use the 'Pooled variance estimate' results. Note that the F-test for equality of variances assumes that the underlying populations are normal; if this is not the case, the F-value maybe suspect as the test is not particularly robust.

In our example, the F-value is not significant, so the 'Pooled variance estimate' results apply. The t-statistic comes to 1.64 and just fails to reach significance at the 10% level, given that we are applying a *two-tailed* test (our hypothesis regarding mean ages was exploratory, remember?). We should therefore conclude that there is no evidence that the mean age of residents in Schweinebratenhausen is different from the mean age of Underwood residents. If we were considering a directional hypothesis, then the p-value would be half that observed in Table 12.7 and significant at $p < 0.10$ (see also discussion on one- versus two-tailed p-values under the one-sample t-test in Chapter 11).

z-test for differences in means

Note that there is an alternative to the two-sample t-test that can be used to test for differences in means between two groups. This is the **z-test for differences in means** (based on the normal distribution) and is discussed in the texts suggested in the Further Reading section. The reasons we are not considering it here are identical to those offered in Chapter 11 for preferring the one-sample t-test over the one-sample z-test (i.e. we usually do not know the population variances; with large samples the t- and z-tests give practically identical results; and, with small samples and/or distinctly non-normal distributions, it is best to opt for a non-parametric test anyway – the Mann–Whitney U test is a good choice in this respect).

Before moving on to greener pastures, we should briefly touch on two issues if only to demonstrate our commitment, our responsibility, our devotion, in fact our reverence, sense of duty and unfailing dedication as educators (sorry, we always get carried away at this point). The first is a rather obvious suggestion concerning the use of the F-test for equality of variances. While we have presented this test purely as a filter for deciding which version of the t-test to use, there is nothing stopping you from using it to test *substantive* hypotheses concerning differences in variability between two populations. Thus you can (and should) use the F-test in a fashion analogous to that of the (single sample) variance test (see Chapter 11).

The second issue is a reminder. Thinking back to Chapter 11, you will recall that the one-sample t-test output contained all the information necessary for calculating a confidence interval for the population mean. The same applies to the two-sample t-test output (see Table 12.7). All *you* need to do is remember that, for example, 90% of the values in the sampling distribution of the mean fall within ± 1.645 standard errors (see Chapter 9, particularly Figure 9.3). Given that both the sample means and the standard errors are provided in the computer output, and assuming that a multiplication with 1.645 is not too much of a burden, you could easily construct confidence intervals for each group; this should further elevate

HINT 12.3

The t-test output makes it easy to calculate confidence intervals.

your status as a top-notch analyst (at least in the eyes of your grandmother!).

One-Way Analysis of Variance (ANOVA)

Having mastered the comparison of two means with the help of the
two-sample *t*-test, you might find the newly discovered analytical power at
your fingertips so tempting that you want to investigate whether three (or
even more) population means are equal. This you can do by running a
one-way analysis of variance; of course, as with a *t*-test, you need interval-
or ratio-level data in order to (meaningfully) calculate means.

The null hypothesis tested by one-way ANOVA is that *k* groups have
equal means in the population ($k \geq 3$); the alternative hypothesis is that *at
least one* mean is different from the others. Note that the alternative
hypothesis does not indicate *which* groups may differ, only that the groups
are not all the same; additional analysis is necessary to identify where the
identified differences exist (we will return to this issue shortly).

Two assumptions must be met before you can legitimately run a one-
way ANOVA. First, each of the groups must be a random sample from a
normal population (note that 'normal' refers to the shape of the distribution not to a state of mind). Second, the variances of the groups must be
equal. As a minimum precaution, you should plot the data for each group
in a histogram (see Chapter 7) and visually inspect whether the distribution is approximately normal. To satisfy the more sophisticated statistical
connoisseur, you could run a one-sample Kolmogorov–Smirnov test to
check whether the normality assumption holds (see Chapter 11). Note
that even if the normality assumption does not *quite* hold, the one-way
ANOVA test still gives reasonably good results. Moreover, if the sample sizes
of the groups are relatively similar, the test is also quite robust against violations of the equal-variances assumption. In the case of major violations
in either assumption, play it safe and use the non-parametric (and more
exotic-sounding) Kruskal–Wallis one-way ANOVA described earlier.

As an example, assume that you have collected data in three countries
(China, Peru and Liechtenstein) in order to compare consumer attitudes
towards Black Forest gâteau (a variable called KUCHEN) on a 10-point
semantic differential scale, ranging from 10 = 'ambrosial' to 1 = 'rotten'
(see Chapter 3 on the format of the semantic differential scale). Your grouping variable is labelled COUNTRY (coded 1 = China, 2 = Peru and 3 =
Liechtenstein) and your null hypothesis is that there is no difference in the
mean attitude scores of the consumers in the three countries. Table 12.8
shows the *SPSS/PC+* output resulting from specifying a one-way ANOVA of
KUCHEN by COUNTRY.

While, at first sight, Table 12.8 appears to be as overwhelming as a
manual describing the inner workings of the Space Shuttle, it is really
straightforward. There are four distinct parts comprising the overall output, namely (a) the results of the one-way ANOVA test itself, (b) descriptive
statistics on each group, (c) a check on the equality of variances assumption, and (d) results of multiple comparison tests. Let's deal with the easy
parts first.

> ### ⚑ WARNING 12.4
>
> Before conducting a one-
> way ANOVA, check that
> the normality and equal-
> variances assumptions are
> (at least approximately)
> satisfied.

Table 12.8 An example of one-way analysis of variance

| Variable | KUCHEN |
| By Variable | COUNTRY |

Analysis of variance

Source	D.F.	Sum of squares	Mean squares	F ratio	F probability
Between groups	2	23.7851	11.8926	7.7501	0.0008
Within groups	86	131.9677	1.5345		
Total	88	155.7528			

Group	Count	Mean	Standard deviation	Standard error	95% Conf. Int. for Mean		
Grp 1	10	2.6000	1.2649	0.4000	1.6951	to	3.5049
Grp 2	64	4.1094	1.1700	0.1462	3.8171	to	4.4016
Grp 3	15	3.3333	1.4960	0.3863	2.5049	to	4.1618
Total	89	3.8090	1.3304	0.1410	3.5287	to	4.0892

Group	Minimum	Maximum
Grp 1	1.0000	5.0000
Grp 2	1.0000	5.0000
Grp 3	1.0000	5.0000
Total	1.0000	5.0000

Tests for homogeneity of variances

Cochrans C = max. variance/sum (variances)	= 0.4298,	$p = 0.275$ (approx.)
Bartlett–Box F	= 0.746,	$p = 0.474$
Maximum variance/minimum variance	= 1.635	

Multiple range test

Scheffe procedure
Ranges for the 0.100 level – 3.08   3.08
The ranges above are table ranges.
The value actually compared with Mean (J) – Mean (I) is
$0.8759 \times$ Range $\times \sqrt{[1/N(I) + 1/N(J)]}$

*(Denotes pairs of groups significantly different at the 0.100 level)

Mean	Group	Grp 1	Grp 3	Grp 2
2.6000	Grp 1			
3.3333	Grp 3			
4.1094	Grp 2	*	*	

The part labelled 'Tests for homogeneity of variances' provides several alternative tests for testing the equality of variances assumption; as long as these are *not* significant (as is the case in our example), we can proceed with our one-way ANOVA (see Warning 12.4 above).

The part of the output immediately preceding the equality of variances test provides a descriptive picture of the groups in terms of group means, standard deviations, standard errors and minimum/maximum values; these summary measures provide us with an overall 'feel' of the data in each group. Moreover, 95% confidence intervals for each group mean are indicated, which give us an idea as to the likely 'overlap' between the three groups.

Now, here's the important part. The key portion of the output is at the very top and shows whether the three groups are different or not. The way that this is done is by *partitioning* the overall variability in the data into two sources: variability of the observations *within* each group (denoted as 'Within groups' in Table 12.8) and variability *between* the group means (denoted as 'Between groups' in Table 12.8). *Total* within-group variability is captured by the **within-group sum of squares** (these are calculated by combining the group variances in a certain way), while a measure of *average* variability within the groups is given by the **within-group mean square** (computed by dividing the sum of squares by the number of degrees of freedom; the latter is equal to $N-k$, where N = *total* sample size and k = number of groups). The corresponding measures for between-groups variability are the **between-groups sum of squares** and the **between-groups mean square** (again the latter is formed by dividing the total sum of squares by the number of degrees of freedom which, in this case, is equal to $k-1$).

within-group sum of squares
within-group mean square

between-groups sum of squares
between-group mean square

If the null hypothesis is true (i.e. there are no differences between the means in the three populations), the variability within each of the three groups should be about the same as the variability between the three groups. Since the within-groups mean square is based on how much the observations vary within each of the groups, while the between-groups mean square is based on how the group means vary among themselves, if the null hypothesis is true, the two mean squares should be similar in value. In other words, if we form the ratio

$$\frac{\text{between-groups mean square}}{\text{within-group mean square}}$$

we should expect it to be close to 1. This ratio is labelled '*F*-Ratio' in the output and its significance can be established by comparing it to the critical values of the *F*-distribution; this is usually done automatically by the computer program (although there are also published tables of the *F*-distribution for those of you using cheap computer software!).

In our example, the *F*-ratio is 7.75 which is not at all close to 1. The associated *p*-value of 0.0008 also signals that it is very unlikely to find such a value if the null hypothesis is true. We therefore reject our null hypothesis and conclude that people in China, Peru and Liechtenstein are likely to have different attitudes towards Black Forest gâteau.

Having unearthed major insights into the culture-bound appreciation of Black Forest gâteau (what a dazzling research topic!), your appetite for pushing the frontiers of knowledge even further will lead you to question *where*, exactly, the differences lie. So far, we only know that the population means are unlikely to be equal. Looking at the descriptive information on the groups, we can see that Peru has, on average, the most favourable attitude, while China has the least. However, we do not know which groups are significantly different from one another. For example, is the difference between Liechtenstein and China statistically significant but that between Liechtenstein and Peru not?

To address these questions we need to look at the final part of the output, the one headed 'Multiple Range Test'. Here we are given the results of what are known as **multiple comparison tests** which pinpoint exactly between which groups the differences exist. In our example, there are significant differences between groups 1 and 2 as well as between 2 and 3, i.e. between China and Peru and between Peru and Liechtenstein; note that there is no difference between groups 1 and 3 (China and Liechtenstein). Armed with this knowledge, we can finally conclude that Peru has a more positive attitude towards Black Forest gâteau than either China or Liechtenstein but no difference in attitude can be identified between the latter two countries.

multiple comparison tests

The particular multiple comparison procedure we applied in our example is called the **Scheffe test**. However, there are several other multiple comparison tests available, with such intriguing names as the Student–Newman–Keuls test (which sounds positively dangerous), the LSD test (which ought to be outlawed) and the Tukey-*b* test (which may remind you of the mating call of a roadrunner). These procedures differ mainly in how they calculate the significance level, taking into account the number of pairwise comparisons. Stick to the commonly used Scheffe and Tukey-*b* tests and you cannot go wrong.

Scheffe test

A final word of warning before we part company with the one-way ANOVA. Do not attempt to run *t*-tests for all possible pairs of means instead of using a multiple comparison procedure. The more means you have to compare, the more likely it becomes that you will find a statistically significant difference even if, in reality, there is no difference between the means in the population. This is the very reason why multiple comparison tests are used – to take into account the fact that many comparisons are made.

WARNING 12.5

Do not replace multiple comparison tests with a series of pairwise *t*-test comparisons. If you do, you will artificially (and wrongly) increase the chance of finding a statistically significant difference in your sample even if there is *no* difference between the means in the population.

Related Measures: Comparing Variables

Sometimes in a data analysis project you may want to undertake comparisons between two or more variables rather than between two or more groups. A typical situation is when you have a research design in which you take some measurements of the same group (i.e. sample of respondents) before and after an event. For example, you may question a set of respondents about their attitude towards a particular product (say, an electric toenail-clipper). Subsequently, you show these individuals an advert featuring the product being used by a celebrity (say, the Queen). Afterwards, you measure their attitude again to find out whether the advert had a positive effect on attitudes. Clearly, in this case, you are dealing with related measures since the same individuals are being compared, albeit on different variables (i.e. pre- and post-advertisement attitudes, respectively).

However, it is not only longitudinal designs (involving taking measurements at different points in time) that result in related measures. You often get the latter in cross-sectional designs as well, for example, when you ask the same group of people to rate different attributes of a certain product and you want to compare the average importance attached to each attribute. Thus you could have asked a sample of consumers to rate three key attributes of an electric toenail-clipper (say energy consumption, adjustable blade and price) on a 5-point importance scale. Subsequently, by comparing the average ratings, you can find out which of the three attributes is considered more important than the others (so that you can emphasize it in your next multi-million advertising campaign). Again, you have a situation where the same group is measured on different variables, that is, you are again facing related measures.

As was the case when comparing groups, when comparing measures, the interest lies in identifying differences that are likely to apply in the *population*. Thus, again, it is important that observed differences based on sample data are subjected to significance tests to ensure that they are not simply the product of sampling error.

Hypotheses involving comparisons between related measures take the following general form:

- There is no difference between the levels of two (or more) measures (null hypothesis).
- There is a difference in the levels of the two (or more) measures (alternative hypothesis – exploratory).
- The level of one measure is higher/lower than that of the other measure(s) (alternative hypothesis – directional).

The overall approach to undertaking comparisons between related measures is very similar to that involving comparisons between groups. Initially, you must establish a (null) hypothesis of no difference. Next, you must select a statistical test which is appropriate for (a) the number of measures to be compared (i.e. two versus more), and (b) the level of measurement involved (i.e. whether you want to compare nominal, ordinal, or interval/ratio measures).

Figure 12.1 (see beginning of the chapter) lists the most important statistical tests for undertaking related measures comparisons. Conceptually, these are very similar to the tests we have just considered for investigating differences between groups; for example, the paired-sample sign test is the equivalent to the Mann–Whitney-U test, while the paired-sample t-test is directly analogous to the two-sample t-test. This simplifies things considerably and allows us to cover the ground more quickly. As we know that your tolerance for learning yet more tests is diminishing at an alarming rate, what we will do is (a) give examples of null hypotheses for each of the tests, (b) use the same data set to test these hypotheses, and (c) present the relevant *SPSS/PC+* output with only the briefest of comments; for further details you should consult our exquisite Further Reading section. Again, for each level of measurement, we will first discuss comparisons between two measures and, while the iron is hot, strike again and extend the discussion to k-measures (we will stick to three measures for illustration purposes); frequent reference to Figure 12.1 should ensure that you do not get lost (or, if you do, that you can find your way back!).

The data set that we will be using in all that follows is based on a survey of 85 randomly-selected shoppers in the island of Lesbos (Greece) who have been asked a number of questions concerning their lifestyles, product preferences and shopping habits. The answers to these questions have been recorded in a variety of ways, yielding nominal, ordinal, as well as interval and ratio data. The overall interest lies in identifying similarities and differences in the patterns of responses by these 85 shoppers and testing to see whether these are likely to hold in the population. Note that the sample size may fluctuate somewhat in the various analyses owing to missing data.

The McNemar Test

McNemar test   The **McNemar test** is very similar to the chi-square test as applied to a 2×2 contingency table, that is it is used when two dichotomous variables are involved. The null hypothesis is that the proportions of subjects with the characteristic of interest is the same under two conditions/treatments.

In our shopper survey, suppose we asked the respondents whether they intended to go to the local cinema during the week (to watch *Last Mango in Paris*) and recorded their answers as 1 = yes, 0 = no; we then gave them a coupon which would entitle them to a free tub of popcorn at the cinema and asked them again whether they intended to go (coding their answers as before, i.e. 1 = yes, 0 = no). The McNemar test allows us to test whether there was any change in their intentions as a result of the coupon; such changes are captured in the upper left and bottom right cells of the 2×2 table, as Table 12.9 shows.

Table 12.9   An example of
the McNemar test

BEFORE with AFTER	Intention before coupon Intention after coupon				
		AFTER			
		1	0	Cases:	84
BEFORE	0	37	10	Chi-square:	20.9302
	1	31	6	Significance:	0.0000

The McNemar test normally uses the familiar chi-square statistic (with one degree of freedom) to test for significance. However, if fewer than 10 cases have different values for the two dichotomous variables, the binomial distribution is used instead. Interpretation of the results is very straightforward: the significant test statistic tells us that there has been a change in the intentions as a result of the free coupon. Looking at the cross-tabulation, we can see that of the 47 shoppers that did not originally intend to go and watch *Last Mango in Paris*, 37 (78.7%) have changed their mind, whereas from those that were intending to go (31 in the first place), only 6

(16.2%) have now decided not to go (probably because they hate coupons, popcorn, or both!).

The Cochran Q Test

The **Cochran Q test** is nothing but an extension of the McNemar test to a situation when three or more dichotomous variables are involved. The null hypothesis is that the proportions of subjects are the same across the set of (three or more) measures.

For example, if we had asked our shoppers to indicate with a yes/no answer (where 1 = yes, 0 = no) whether they like (a) oysters, (b) prawns, and (c) crabs, we could use the Cochran Q test to test for equality of preferences across the three types of seafood. The output we would get is shown in Table 12.10.

Cochran Q test

	Cases			
0.00		1.00	Variable	
22		61	OYSTER	Like oysters
16		67	PRAWN	Like prawns
47		36	CRAB	Like crabs

Cases	Cochran Q	D.F.	Significance
83	28.4561	2	0.0000

Table 12.10   An example of the Cochran Q test

The Q-statistic follows approximately a chi-square distribution with $k–1$ degrees of freedom, where k = number of variables compared (here $k = 3$, so DF = 2). The significant result indicates that we have to reject the hypothesis of equal preferences; indeed, looking at the displayed frequencies we can see that the preferences for crabs are considerably different from those relating to oysters and prawns. For some reason – which only need concern owners of seafood restaurants – only about 40% of the sample like crabs as compared with 73% for oysters and 80% for prawns.

The Paired-Sample Sign Test

If you want to compare two ordinal measures, this is a convenient test to use. Casting your mind back to Chapter 11, you may recall that one of the location tests we considered was the one-sample sign test (we used that to test hypotheses concerning a median). The **paired-sample sign test** works in a very similar way: it tests the null hypothesis that the median *difference* of two measures is zero.

Table 12.11 shows the results of a paired-sample sign test applied to our shopper survey; specifically, respondents were asked to indicate on a 5-point scale how often they go to the opera and theatre, respectively (the scaling format used was as follows: 5 = weekly, 4 = fortnightly, 3 = monthly, 2 = annually, and 1 = every seven years). The question of interest is

paired-sample sign test

Table 12.11   An example of the paired-sample sign test

OPERA with THEATRE	Opera attendance Theatre attendance	
Cases		
35	− Diffs (THEATRE Lt OPERA)	$z = 2.8571$
14	+ Diffs (THEATRE Gt OPERA)	
30	Ties	2-tailed $p = 0.0043$
—		
79	Total	

Lt = less than, Gt = greater than

whether the median attendance levels are the same for opera and theatre or not.

The test statistic is based on the normal distribution (i.e. a z-value is produced) as long as there is a reasonable number of cases; if the sample size is less than 30, the test uses the binomial distribution (this also applies to the one-sample sign test – see Tables 11.8 and 11.9 in Chapter 11). The significant result indicates that the median difference is unlikely to be zero, that is opera and theatre attendance are not equally frequent. In fact, looking at the differences between the pairwise ratings, the greater number of instances favouring opera indicates that attendance for the latter is higher than for theatre (for more details on the rationale of the test and the interpretation of the signed differences, we refer you back to its single-sample equivalent in Chapter 11).

The Friedman Two-Way Analysis of Variance (ANOVA)

Friedman two-way analysis of variance

The **Friedman two-way analysis of variance** is the appropriate test to use when several ordinal-level measures need to be compared to one another; conceptually, it is very similar to the Cochran Q test discussed earlier, the only difference being that instead of dichotomous we have rank order data. The null hypothesis is, of course, that there are no differences among the set of measures.

Let us assume that we have asked our sample of shoppers to rank three local supermarket chains – *Superbuy*, *Extrasave*, and *Megadeal* – in order of preference (1 = most preferred, 3 = least preferred) and we are interested in finding out whether the three chains are equally preferred. Table 12.12 shows the results of applying a Friedman two-way ANOVA to the responses obtained.

Again, the test statistic follows an approximation of the chi-square distribution with $k–1$ degrees of freedom (where k = number of variables); the significant result obtained indicates that preferences among shoppers for the three supermarkets are not equal. If the sample size n and the number of variables k are very small, the test statistic is not exactly distributed as a chi-square, but there are published tables for it for different combinations of k and n (such tables can be found in the texts listed in the Further Reading section).

Mean rank	Variable
2.27	SUPERBUY
1.95	EXTRASAVE
1.78	MEGADEAL

Cases	Chi-square	D.F.	Significance
79	9.9304	2	0.0070

Table 12.12 An example of Friedman's two-way analysis of variance

The Paired-Sample *t*-test

This test is the related measures equivalent to the two-sample *t*-test for differences in means (it is also known as the 't_r-test' to distinguish it from the 'conventional' *t*-test). It lends itself nicely to comparisons of two interval or ratio-level measures, the null hypothesis being that the mean difference in the population is zero.

Table 12.13 shows the results of a **paired-sample *t*-test** aiming at identifying whether there is a difference in the quality of service provided by two different restaurants (*Nick's Garden Dungeon* and *Tony's Jolly Hellhole*), as perceived by our sample of shoppers. A 5-point semantic differential scale ranging from 1 = appalling to 5 = outstanding was used to register opinions. Those of you with an incredible memory will have noticed that the format of the output is very similar to that of the one-sample *t*-test (see Table 11.11 in Chapter 11).

paired sample *t*-test

Variable	Number of cases	Mean	Standard deviation	Standard error
NICK	79	3.2658	1.184	0.133
TONY	79	2.8228	0.902	0.102

(Difference) Mean	Standard deviation	Standard error	*t*-value	Degrees of freedom	2-tail probability
0.4430	1.196	0.135	3.29	78	0.001

Table 12.13 An example of the paired-sample *t*-test

You get all sorts of goodies from the output of the paired-sample *t*-test: the means and standard deviations of the two variables, their standard errors (so you can, yet again, calculate confidence intervals) as well as the difference between the means and its standard error. The *t*-value and associated two-tailed probability are also shown.

The significant test statistic indicates that the service quality evaluations of the two restaurants are not the same. Indeed, judging from the mean values, *Nick's Garden Dungeon* is more highly rated than *Tony's Jolly Hellhole* (having said that, as neither rating is particularly high, the quality of service in both 'establishments' may leave something to be desired!).

The t_r-Test for All Pairs

t_r-test for all pairs

The **t_r-test for all pairs** is simply the application of the paired-sample *t*-test to a situation where more than two variables are involved. For example, if our sample were asked to rate *George's Wacky Tavern* alongside *Nick's Garden Dungeon* and *Tony's Jolly Hellhole*, we would make *three* paired comparisons of service quality ratings. In other words, we would apply the paired-sample *t*-test three times (taking two restaurants at a time) and, depending upon which particular pairs were found to differ, we would draw conclusions concerning the existence and nature of differences between the three restaurants. It's as simple as that.

A final word concerning the comparison of related measures: the comparison undertaken must make *conceptual* sense. For example, comparing the importance of price as a choice criterion when buying a car with the importance of reliability when buying a washing machine is not a good idea. Even if the same scale is used in both instances to measure importance, the comparison would be nonsensical – so *think* before you compare.

Summary

This chapter introduced you to the undoubtedly pleasurable activity of making comparisons. We first reminded you of the difference between independent and related measures, as different statistical techniques are involved in each case. We then looked, in some detail, at different tests for undertaking comparisons between groups and followed this with a discussion of the most common tests for comparing sets of measures. You should now be in a position to select the right technique to carry out comparisons, given the number of groups/measures involved and the level of measurement of the variables. If you are still feeling a bit uncertain, just refer back to Figure 12.1 for a quick confidence booster.

Questions and Problems

1. List the considerations you have to go through to decide on the appropriate statistical technique for making comparisons.

2. You are comparing two samples of male and female train passengers on their drinking habits. The passengers drink either tea, coffee, vodka, or beer. Which test would you use to compare males with females in this setting?

3. Explain under which circumstances you would use the Fisher exact test.

4. Is it possible to apply test statistics used for higher levels of measurements to situations involving lower levels of measurement?

5. You have devised a service quality measure for waiters at Moscow Intourist Hotels. It is a three-point scale: 'none', 'lousy', 'rudimentary'. If you intended to compare a sample of 35 waiters over 40 years of age with a sample of 27 waiters who are all below 40, which statistical test would you use?

6. You have obtained samples of 100 railway workers each in Nepal and Ireland, and you intend to compare them on the number of cigarettes smoked during the week prior to sampling. Which statistical test are you likely to use?

7. Construct an example built around samples of three different groups of workers employed by the Belgian Ministry of Agriculture, which would enable you to run a Kruskal–Wallis one-way ANOVA.

8. Explain when you would use (a) a paired-samples *t*-test, (b) Friedman's two-way ANOVA, and (c) Cochran's Q test.

9. What are multiple comparison procedures? With which statistical technique are they associated?

10. How does this chapter compare to Chapter 11 in terms of entertainment value?

Notes

1. Norusis, MJ (1991) *The SPSS GUIDE to Data Analysis for SPSS/PC+*, 2nd edn, p. 271. Chicago: SPSS Inc.

Further Reading

Babbie, E & Halley, F (1995) *Adventures in Social Research: Data Analysis Using SPSS for Windows*. Thousand Oaks, CA: Pine Forge Press. This is a hands-on guide for doing data analysis with the *Windows* Version of *SPSS*; a major data set is used throughout to illustrate the application of the various techniques and several computer screens are also included. Useful.

Norusis, MJ (1991) *The SPSS GUIDE to Data Analysis for SPSS/PC+*, 2nd edn. Chicago: SPSS Inc. An excellent guide to data analysis, very practice-oriented and to the point. If you use *SPSS/PC+* as your analysis package, this book is a must.

Sirkin, RM (1994) *Statistics for the Social Sciences*. London: Sage Publications. A comprehensive textbook on statistical analysis covering most of the techniques discussed in this and the next chapter.

13

Getting adventurous: searching for *relationships*

The Mystique of Relationships

Relationships with different people form the fundamentals of our lives. There are relationships, for example, between mother and child, between colleagues at work, between lovers, and even with the friendly public toilet attendant who always acknowledges 'regular' customers with a friendly nod but whose name one does not know. Of course, the nature of these relationships varies considerably as does their strength: some are quite casual while others are very intense.

Relationships (or 'associations') also play an important role in data analysis. Very often we want to find out whether two variables are related and, if so, the nature and strength of this relationship. To do this, we employ what are known as **measures of association** and we will consider a few of them later in this chapter. Note that our focus will be on **bivariate relationships** only (i.e. those involving two variables) and, for the time being, we will conveniently ignore **multivariate relationships** (i.e. those involving three or more variables); however, we will give you a quick taste of multivariate analysis in the next chapter.

In studying relationships, we frequently ask questions, such as: Is advertising related to sales? Is the number of children related to happiness? Is the number of Royal Family scandals related to newspaper circulation? Questions of this kind are primarily directed at discovering *whether* a relationship exists between variables. However, in many cases, we are not only interested in the mere existence of a relationship, but also in the way in which the two variables are related to one another, that is, in the *direction* of the relationship. For example, we may ask: Is a decrease in price associated with an increase in sales? Is hair loss positively related to beer drinking? Is there a negative link between church attendance and attendance at rave parties?

Another aspect of a relationship we may be interested in is its *magnitude*. Thus, although we may find, say, a positive relationship between advertising and sales, this relationship may not be as strong as, say, between price reductions and sales. The magnitude (or 'strength') of a relationship tells us how closely two variables are related to one another. In this context, measures of association are usually calibrated so as to range between 0 and ±1, with 0 indicating no relationship between the variables (i.e. com-

(margin notes)

measures of association

bivariate relationships
multivariate relationships

plete independence) and 1 a perfect relationship (the (+) or (−) sign indicating the direction of the relationship); intermediate values reflect different magnitudes of the relationship. As a very rough rule of thumb, a relationship is usually considered 'strong' if the association measure is larger than 0.8. Between 0.4 and 0.8 we are in the 'moderate' region and below 0.4, the relationship tends to be considered as 'weak'. Note that, with real-life data, it is very unlikely that a measure of association will reach its extreme values (i.e. 0 or 1). For example, even if in the population the two variables are totally unrelated, a measure of association will generally produce a non-zero value owing to sampling error. Thus, when dealing with sample data, it is important to *test* whether a value produced by a measure of association does in fact reflect the existence of a 'true' relationship in the population. This is easily done as most measures of association are usually accompanied by a significance test which tests whether the association between the variables in the population is indeed significantly different from zero.

WARNING 13.1

Always accompany a measure of association with a statement regarding its significance.

Bearing all the above in mind, hypotheses involving relationships between two variables X and Y can be stated as follows:

- There is no relationship (i.e. zero association) between the two variables: X and Y are unrelated (null hypothesis).
- There is a relationship (i.e. non-zero association) between the two variables: X and Y are related (alternative hypothesis – exploratory).
- There is a positive/negative relationship between the two variables: X is positively/negatively related to Y (alternative hypothesis – directional).

Measures of Association

There is a large number of (bivariate) measures of association that can be used to examine relationships between variables (see Further Reading section for examples). After careful deliberation, many sleepless nights, close consultation with 43 Nobel laureates in statistics, and three visits to a fortune-teller, we selected three of them (which also happen to be the most widely used ones): **Cramer's V** for looking at the association between nominal variables; **Spearman's rank-order correlation** for capturing the relationship between ordinal variables; and **Pearson's product moment correlation** for examining linkages between interval- and/or ratio-scaled variables. We will consider each of them in turn and also mention in passing some other association measures.

Cramer's V
Spearman's rank-order correlation
Pearson's product moment correlation

Cramer's V

You may recall from Chapter 10 that statements concerning relationships between nominal variables, such as hair colour, religious denomination or nationality, are inherently limited by the nature of the measures. Specifically, it does not make sense to try to interpret the *direction* of an association between nominally-scaled variables, as demonstrated by senseless statements like 'as nationality increases, religious denomination

WARNING 13.2

Do not attempt to interpret the direction of a relationship between nominal variables.

WARNING 13.3

The chi-square statistic can only establish whether two nominal variables are independent or not. It does *not* show the strength of the association between the variables.

decreases' or 'hair colour is negatively related to religious denomination'. Since categories of nominal variables do not possess a meaningful order (e.g. blond is not better or worse than black), they cannot be related in a particular direction. Indeed, given the arbitrary scoring of nominal scales (see Chapter 3), whether a relationship turns out to be positive or negative depends purely on how categories are scored rather than on any intrinsic pattern in the data.

However, what you can interpret is the *strength* of the relationship between the two variables and this is where Cramer's V comes in handy. To understand Cramer's V, it is necessary to recall our discussion of the chi-square statistic in Chapter 12. You may vaguely remember that we used chi-square to test the null hypothesis that there is no difference between the actual and expected frequencies in each cell of a contingency table, i.e. that the two nominal variables forming the table are independent. Unfortunately, even if we reject the hypothesis of independence, we cannot conclude anything about the strength of the relationship between the two variables. This is because the value of the chi-square statistic is dependent on the sample size and the size of the contingency table; as a result, the value of chi-square cannot be relied upon to distinguish between different relationships.

But why spend time demonstrating which technique does *not* work instead of telling you one that *does*? And why do marmalade sandwiches always land on the buttered side when they fall off the table? While we have still not solved the last problem, the answer to the first question is that chi-square provides the basis for many measures of association developed by statisticians. While, on its own, chi-square can only test independence, it can be modified so that (a) it is not influenced by sample size, and (b) its values fall in a range from 0 to 1 (where 0 indicates no association and 1 perfect association).

Cramer's V represents such a chi-square-based adjustment. Its values always fall between 0 and 1 and, thus, can be interpreted as reflecting relationships of different magnitudes. Cramer's V is often provided as a statistical option in computerized analysis packages; if your package does not include it, you can easily calculate it by hand if you know the chi-square value. The relevant formula is as follows:

$$V = \sqrt{\frac{\chi^2}{n(k-1)}}$$

All you have to do is divide the chi-square value by n (the sample size) times $(k-1)$, where k is the *smaller* of the number of rows or columns in the contingency table, and then take the square root. If either the number of rows or columns equals 2, it follows that you only need to divide the chi-square value by the sample size. Statistical connoisseurs (should one have accidentally picked up this book) will notice that, in this special case, Cramer's V is identical to the so-called **phi coefficient** (which is a measure of association for 2×2 tables, i.e. when two dichotomous variables are related to one another). While this is completely inconsequential for most of us, we thought we ought to mention it in a desperate attempt to raise the academic standard of the book!

phi coefficient

To show you how easy it is to calculate Cramer's *V*, we will use the data in Table 12.1 in Chapter 12 (and we categorically dismiss any suggestion that the 'recycling' of Table 12.1 reflects laziness on our part to come up with a fresh example!). Just in case you cannot recall the relevant information from memory, the chi-square value was 1.08, the table had three rows and two columns, and the sample size was 1461. Plugging this information in the above formula, we get the following:

$$V = \sqrt{\frac{1.08}{1461\,(2-1)}} = 0.027$$

This is as close to zero as you can get, so we can be fairly sure that there is no relationship whatsoever between country of origin and preferred headache remedies. In any case, the fact that the chi-square statistic was not even remotely significant in the first place (its *p*-value was 0.582) would have warned us not to raise our hopes when looking at the strength of the relationship (because there is none!). In this context, you should look at and report both the chi-square statistic *and* Cramer's *V*, as the interpretation of the latter rests on your ability to reject the null hypothesis of independence between the two nominal variables (as tested by the chi-square).

It is worth noting that an advantage of Cramer's *V* is that it enables us to compare contingency tables of different sizes and, equally importantly, tables based on different sample sizes; the chi-square statistic allows us to do neither.

HINT 13.1

Report both the chi-square and Cramer's *V* statistics to make interpretation clearer.

Spearman's Rank-Order Correlation

If you are dealing with a situation in which both variables concerned are ordinal, you can investigate not only the strength of the association but also its direction (and, thus, distinguish between positive and negative relationships). Spearman's rank-order correlation coefficient (also sometimes referred to as 'Spearman's rho') is an appropriate measure of association in this case. It ranges from −1 to +1, with values close to zero indicating little or no association between the variables concerned. Moreover, its sampling distribution under the null hypothesis is known (it follows approximately a *t*-distribution), so we can test for significance.

Consider the following example for further enlightenment. We are trying to find out whether there is a relationship between liking the Royal Family and liking these strange little dogs (known as corgies) Her Majesty the Queen is often seen with. To do this, we asked a random sample of 145 Japanese golfers to indicate their liking for the Royal Family on the following scale: 1 = 'I don't like them', 2 = 'they are OK – just', and 3 = 'I would like to move in with them'. Liking for the corgies was captured by responses to the following scale: 1 = 'would not waste a photo on such a dog', 2 = 'would only take a photo from a safe distance', 3 = 'would like to take lots of photos', and 4 = 'would like to be photographed with such a dog on my lap'. Using the by now rather familiar cross-tabulation (as produced by *SPSS/PC+*), the data is summarized in Table 13.1.

Table 13.1   Liking the Royal Family and liking corgies

ROYALPREF BY DOGPREF

		DOGPREF				
	Count	No photo	Distant photo	Lots of photos	Joint photo	Row total
ROYALPREF		1	2	3	4	
I don't like them	1	27	10	5	5	47
They are OK – just	2	4	5	16	31	56
I would like to move in with them	3	9	18	7	8	42
	Column total	40	33	28	44	145

Statistic	Value	ASE1	*t*-value	Approximate significance
Spearman correlation	0.22552	0.08627	2.76819	0.005

Requesting the Spearman rank-order correlation coefficient, we obtain a value of 0.226, with a standard error estimate of 0.086 (denoted as 'ASE1' in the output). The corresponding *t*-value comes to 2.77, which is highly significant. This means that the observed correlation is unlikely to have come about *if* there was no association between the two variables in the population (i.e. if the population correlation coefficient was equal to zero). Thus, we can reject the null hypothesis that there is no relationship between liking the Royal Family and liking corgies. In fact, according to our findings, the two variables are positively related, although the strength of the relationship is rather weak (as indicated by the magnitude of the coefficient).

Kendall's rank-order correlation

Note that we would have reached exactly the same conclusion had we employed **Kendall's rank-order correlation** (also known as 'Kendall's tau-c'), which is another widely used measure of association for ordinal variables (it would have produced a value of 0.187, which is also significant at $p < 0.01$). Although there are some technical differences between Spearman's and Kendall's coefficients, there is little to distinguish between them in practical applications: use whichever makes you happy (or whichever you can get from your analysis package). Having said that, you will have noticed that the values for Kendall's and Spearman's rank-order correlation coefficients are different (0.226 versus 0.187), although we used the same data in both cases. The reason for this is that different algorithms are used in their calculation. Consequently, if we measured the association between two different sets of variables and used Spearman's coefficient in one case and Kendall's coefficient in the other, we could not compare the results. However, if applied to the same set of data, both measures will reject the null hypothesis at the same level of significance (as was the case in our example).

WARNING 13.4

Don't 'mix and match' Spearman's and Kendall's rank-order correlation coefficients; numerically, they are not directly comparable to each other.

Pearson's Product Moment Correlation

This is the most widely used measure of association for examining relationships between interval and/or ratio variables. Also known as 'Pearson's *r*', the product moment correlation coefficient focuses specifically on **linear relationships** and ranges from −1 (a perfect negative linear relationship), through 0 (no linear relationship) to +1 (a perfect positive linear relationship). The emphasis on 'linear' is important because if two variables are linked to one another by means of a **nonlinear relationship** the Pearson correlation coefficient cannot detect it. For example, calculating Pearson's correlation coefficient based on the data displayed in Figure 13.1 is likely to produce a small and non-significant value; this would suggest no relationship between the variables despite the fact that a substantial (albeit curvilinear) relationship exists between the two variables.

linear relationship

nonlinear relationship

Figure 13.1 An example of a nonlinear relationship.

[Scattergram: Variable A (y-axis, 0 to 700) vs Variable B (x-axis, 0 to 120), showing an inverted-U shaped distribution of data points.]

To avoid such pitfalls, it is always a good idea to plot the relationship between the variables in a **scattergram** (or 'scatter plot') before applying Pearson's correlation coefficient.

Figure 13.2 shows a scattergram for what is clearly a linear relationship between the sales for left-handed tea cups in a sample of export markets served by *Global Teapot Provisions Ltd* (a manufacturer of teapots, cups, and assorted products) and the number of sales representatives appointed in these markets. A visual inspection of the plot suggests that an increase in the number of sales representatives tends to be associated with an increase in sales and vice versa.

If we now kindly ask our statistical package for the Pearson product moment correlation coefficient, we should get something along the lines of Table 13.2. The coefficient is positive, of substantial magnitude ($r = 0.75$) and highly significant ($p = 0.000$); with regard to the latter, you should bear in mind (as mentioned previously) that the three zeros arose from rounding - the exact probability may be, say, $p = 0.00003$. Staying with the *p*-value, you should further notice that one can usually choose

scattergram

HINT 13.2

Plot your data whenever possible. A visual inspection helps to gain a better understanding of the relationship between two variables.

Figure 13.2 An example of a linear relationship. (# indicates two or more cases share the same values.)

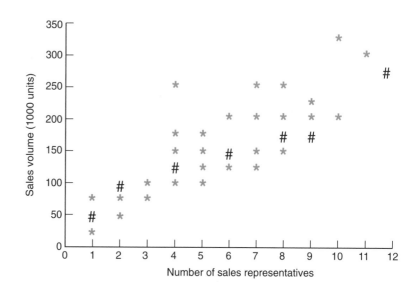

Table 13.2 An example of Pearson's product moment correlation

	SALESVOL
	0.7517
SALESREP	(45)
	$p = 0.000$

(Coefficient/(Cases)/1-tailed significance)

HINT 13.3

Use a one-tailed significance test of the correlation coefficient if you know *in advance* the directionality of the relationship between the variables. Otherwise use a two-tailed test.

proportion of variance

correlation matrices

whether the computer package reports one-tailed or two-tailed significance levels. Report the one-tailed probability whenever you are confident about the direction of the relationship (whether positive or negative). In cases where you do not know the directionality of the expected relationship, you should report the two-tailed *p*-value (see also earlier discussion on choosing one- versus two-tailed test in Chapter 10).

A Pearson correlation coefficient of 0.75 tends to suggest a 'moderate to strong' association between the two variables. However, we can go beyond this somewhat imprecise description and actually calculate the **proportion of variance** in sales volume that is 'explained' by the number of sales representatives. This is easily achieved by simply squaring the correlation coefficient, i.e. by calculating r^2. In our example, $r^2 = (0.75)^2 = 0.56$, which indicates that 56% of the variation in sales volume can be explained by the variation in the number of sales representatives. Both r and r^2 measure the strength of a linear association between two interval or ratio-scaled variables. Using r^2 often aids interpretation as it shows the proportion of variance in one variable explained by the other; however, it tells us nothing about the direction of the relationship.

While we pledged to look only at two variables at a time, we would neglect our duties (and, no doubt, upset book reviewers), if we failed to mention one of the most widely used applications of the Pearson correlation coefficient, namely its use in creating **correlation matrices**. Using simple commands such as 'PEARSON CORR Var1 to Var395', statistics programs will readily correlate *all* listed variables with each other (taking two at a time) and indicate which of the resulting relationships are statistically significant. Table 13.3 shows an example of such a correlation matrix for five variables.

Correlations	VAR1	VAR2	VAR3	VAR4	VAR5
VAR1	1.0000	−0.2066	−0.4310*	−0.1284	0.0099
VAR2	−0.2066	1.0000	0.4881*	0.7517**	0.4369*
VAR3	−0.4310*	0.4881*	1.0000	0.5444**	0.1470
VAR4	−0.1284	0.7517**	0.5444**	1.0000	0.5552**
VAR5	0.0099	0.4369*	0.1470	0.5552**	1.0000

Number of cases: 42   2-tailed significance: * 0.01   ** 0.001

Table 13.3 An example of a correlation matrix

Note that all correlations on the diagonal are equal to 1; this should not surprise you, since the correlation of a variable with itself *has* to be perfect. As far as the off-diagonal elements are concerned, note that the correlations forming the two triangles above and below the diagonal are identical; thus, it is only necessary (and customary) to report the correlation results in one of the two triangles and leave the other blank.

Although the ease of producing a correlation matrix tempts many researchers – especially the inexperienced and the lazy – to correlate *all* variables in their study with each other in order to uncover all possible relationships (or, expressed less charitably, to desperately fish for results), we advise you to use this procedure with care. First, such an indiscriminate approach is a giveaway that the study lacks conceptualization and shows that the researchers do not really know what they are looking for. Second, if you compute a large enough number of correlation coefficients, some relationships will turn out to be significant by pure chance. For example, if you compute 100 correlation coefficients, you could expect about 10 of them to show significance levels at $p < 0.10$ or better, even when there is no relationship among these variables in the population. Third, the amount of computer printout that you will generate will be so overwhelming that you will most probably have to spend all your weekends, public holidays, and entire summer vacation poring over the output! To put it in perspective, correlating 4 variables with each other results in 6 distinct correlations, while intercorrelating 8 variables results in no fewer than 28 correlations (in general, given k variables the number of distinct correlations is equal to $k(k–1)/2$ – not a pretty thought!).

Two final points concerning Pearson's correlation coefficient. The first is that it assumes that the *joint* distribution of the variables in the population is normal, i.e. that we are sampling from a **bivariate normal distribution**. Note that this assumption affects the significance test of r, not the computation of r itself; in this context, 'if . . . the joint distribution in the population is not normal, we cannot make inferences about it. However, we can compute descriptive measures of correlation for the sample'.[1] While there are ways of testing for bivariate normality, the procedures involved are way outside the scope of this book. For practical purposes, as long as the *individual* distributions are not markedly non-normal

HINT 13.4

If you are primarily interested in the *direction* of a (linear) relationship use the Pearson correlation coefficient (r). If your main concern is the *strength* of the relationship focus on r^2.

WARNING 13.5

Do not blindly correlate all variables with each other – correlation matrices are always easy to produce but often difficult to justify and interpret!

bivariate normal distribution

and the sample has a reasonable size, one can safely use Pearson's correlation coefficient (if either of these conditions does not apply, one might just as well rely on non-parametric measures of association such as Spearman's and Kendall's rank-order correlations).

The second point has to do with interpretation. Having obtained a statistically significant result, you should pause and think about the *substantive* significance and the limitations of the findings (this is also a good excuse to have a break during which to think about the meaning of life in general). Specifically, you should always keep in the back of your mind that, for very large sample sizes, even small correlation coefficients (e.g. 0.1 or 0.2) may be significant (see also discussion of power in Chapter 10). While such a result would, indeed, indicate a very small linear relationship between the variables, the relationship may not be particularly important from a substantive point of view; indeed, taking an r^2 perspective, only 1% and 4% of variance is explained by the above relationships. Thus, you should consider the 'practical' significance of your correlation coefficient alongside its statistical significance.

WARNING 13.6

Always think about the substantive significance of any statistically significant correlation coefficient.

Correlation and Causality

Whenever you are examining a relationship between two variables, there is always a temptation to draw *causal* inferences on the basis of correlation results. This temptation must be firmly resisted (as should all temptations – well, most of them!). For example, think about the link between sales and advertising. Most people would naturally assume that sales are caused by advertising. Sounds perfectly reasonable until you find out that many companies determine their advertising budget based on last year's sales figures – oops!

The fact that two variables are related, the fact that this relationship can be captured by an association measure, and the fact that this association measure may generate a statistically significant result is no evidence whatsoever that one variable causes the other. No matter how 'intuitively appealing' a cause-and-effect explanation may be and no matter how 'obvious' the designation of each variable as a cause or effect, the fact remains: *correlation does not prove causality*. All that is expressed by an association measure is the degree of covariation between two variables. Any notion of causality must come from practical knowledge or theoretical insights into the subject area, preferably supported by longitudinal data obtained under experimental conditions (the need for longitudinal data becomes evident when you think that a cause *must* temporally precede an effect, while experimental conditions reflect a need to control for other variables so as to 'isolate' the effect from the influence of unwanted causes). While a full-blown discourse on the conditions necessary to enable the drawing of causal inferences would take us into the realm of the philosophy of science, you should think twice before you present correlations among variables as reflecting causal processes.

WARNING 13.7

Correlation results on their own can never prove causality. Interpreting correlation results as causal relationships is erroneous and misleading.

Lastly, some of you may be wondering what to do if you need to investigate the association between variables which have *different* levels of measurement. The techniques discussed in this chapter assume that both variables have the same level of measurement (i.e. they are both nominal, ordinal or interval/ratio). What if the variables are 'mixed'? The solution is simple: if one of the variables is nominal, it is best to conceive the problem as one involving *comparisons* rather than associations, with the nominal variable defining group membership. You can use the various techniques discussed in Chapter 12 for undertaking group comparisons (see in particular Figure 12.1). If one of the variables is ordinal and the other is interval/ratio, then use a non-parametric association measure such as Kendall's or Spearman's rank-order correlation. We should point out that there are specialized measures of association for gauging the strength of relationships between variables measured at different levels of measurement (such as 'polychoric' and 'polyserial' correlations); however, for most data analysis projects, the techniques described in this chapter and Chapter 12 are quite sufficient for dealing with any combination of variables. If you are still skeptical, here's a challenge for you: select any two variables of your choice, have them measured any way you like, and we *bet* with you that somewhere in Chapter 12 or 13 (this one!) you will find a method for analysing the linkage between them.

Summary

In this chapter we focused on uncovering and assessing relationships between two variables. We first distinguished between the existence, direction and strength of a relationship and indicated the importance of accompanying measures of association with a significance test. We then considered several measures of association, with the level of measurement being, again, a key factor determining the correct choice of technique. Finally, we warned against interpreting correlation results in causal terms and gave you some hints on how to deal with situations where the level of measurement is not the same for each variable.

Questions and Problems

1. What is the difference between the strength and the direction of a relationship?

2. Can the chi-square statistic be used to measure the strength of a relationship between two nominal variables?

3. Which values can be assumed by Cramer's V and how should these numbers be interpreted?

4. You have taken a sample of 178 Russian soldiers to establish whether there is a relationship between heroism and flying Aeroflot. Both heroism and the likelihood of flying Aeroflot have been measured on three-point ordinal scales. Which

statistical technique would you use to test for association between these two variables?

5. What is the null hypothesis under the Pearson correlation coefficient? What would a coefficient of 0.9 indicate?

6. Does a Pearson correlation coefficient close to zero *necessarily* imply no association between the two variables under consideration?

7. Under what circumstances would you use a one-tailed test to test the significance of a correlation coefficient?

8. What does the square of Pearson's correlation coefficient (r^2) tell us?

9. Why can we not interpret significant correlation results as indicating causal relationships?

10. Why does this chapter have such an unlucky number?

Notes

1. Daniel, WW & Terrell, JC (1995) *Business Statistics for Management and Economics*, 7th edn, p. 522. Boston: Houghton Mifflin.

Further Reading

Gibbons, JD (1993) *Nonparametric Measures of Association*. London: Sage Publications. Excellent source of reference on measures of association for non-metric data.

Hildebrand, DK, Laing, JD & Rosenthal, H (1977) *Analysis of Ordinal Data*. London: Sage Publications. As Reynolds (1977) but focusing on relationships among ordinal variables.

Reynolds, HT (1977) Analysis of Nominal Data. London: Sage Publications. Good treatment of techniques for analysing linkages between variables measured with nominal scales.

Chapter 4

Getting hooked: a look into a *multivariate future*

The Nature of Multivariate Analysis

This chapter is devoted to the connoisseurs of data analysis; those of you who have truly lost your fear of the subject; those of you who dream of becoming virtuosos of number crunching; those of you who want to elevate the art of data analysis to new and dizzy heights; and, finally, those of you who intend to marry a statistician! In short, it is written for the two per cent of our readership who are interested in learning something about the techniques available for analysing several variables simultaneously, i.e. about **multivariate analysis**. Of course, a single chapter can only provide a very superficial taster of what is a very useful but admittedly rather complex topic (we also realize that the sudden mention of the term 'multivariate analysis' can result in nausea, stomach colic, and uncontrollable eye-twitching among uninitiated novices). At best, we hope to whet your appetite sufficiently for you to borrow/buy a book on the subject and learn more about multivariate analysis. At worst, you should recognize some inherent limitations of the techniques described earlier in detail in this book, realize what other types of analysis you could potentially do, and learn a few fancy terms to impress people at the next Tupperware party you attend!

So what *is* multivariate analysis? Well, let us start by saying that none of the statistical techniques we have so far discussed would qualify for the 'multivariate' title. Up until now, the techniques we have covered were concerned with either the analysis of one variable at a time (for example, a one-sample *t*-test) or the link between two variables at a time (for example, correlation analysis). In other words, we have been discussing techniques suitable for undertaking univariate and bivariate analysis but, alas, not for multivariate analysis.

Multivariate analysis deals with more than two variables *simultaneously*; for example, it enables the comparison of several groups in terms of several variables and the investigation of interrelationships among sets of variables. Multivariate analysis techniques are, in many instances, clear extensions of univariate and bivariate techniques. For example, **multivariate analysis of variance (MANOVA)** enables the comparison of several groups in terms of *multiple* (interval-scaled) variables; thus it can be conceived as an extension of the analysis of variance (ANOVA) procedure we discussed in Chapter 12 (in which three or more groups were compared on

multivariate analysis

multivariate analysis of variance (MANOVA)

a single interval-scale variable). Indeed, the null hypotheses of the two techniques are quite similar: whereas in one-way ANOVA the null hypothesis is that there is no difference in the group means for the variable of interest, in MANOVA the null hypothesis is that there is no difference in the *sets* of means across the groups (since several variables are simultaneously compared). The latter is known as a **multivariate hypothesis**, as its rejection or non-rejection refers to the set of variables *as a whole* rather than to any of the individual variables. To test multivariate hypotheses, **multivariate significance tests** are employed, of which there are several (but which, you will be thrilled to know, we will not attempt to discuss here).

multivariate hypothesis

multivariate significance tests

But what is the benefit of undertaking multivariate analysis? Given that multivariate analysis techniques are substantially more complex than their univariate and bivariate counterparts (as well as more computationally demanding), why not just repeat, say, a univariate analysis a number of times until all variables of interest have been considered? Why should we look at all the variables simultaneously? The answer to this question is best given with an example.

Assume that you are a business analyst and you want to find out the key factors that have an influence on the sales revenue of your firm (a manufacturer of horse blankets); the motivation behind your analysis is to generate some information which you can use to decide how best to allocate your marketing budget among advertising, in-store displays, personal selling, etc. Say also that, on the basis of historical data contained in company records, you have found that the (Pearson) correlation between sales revenue and the amount spent on advertising comes to 0.48 and the correlation between revenue and number of sales personnel comes to 0.56. What can you conclude from this analysis, other than that both marketing variables appear to have a positive influence on revenue? Can you identify the *relative* importance of each marketing variable on sales revenue? Can you say something about the *joint* (i.e. total or combined) impact of advertising and personal selling on revenue?

Unfortunately, the answer to both questions is no. Simple (i.e. bivariate) correlations do not permit us to deal with questions of this sort. While you may vehemently protest at this stage, since it is obvious that the number of sales personnel has a stronger correlation with revenue than does advertising (0.56 versus 0.48), the inference that the influence of the sales personnel is more important is only correct if there is no correlation between it and the amount spent on advertising (i.e. that the two marketing variables are totally independent). In reality, however, this is highly unlikely to be the case (e.g. the decision to use more/fewer salespeople is often accompanied by a reduction/increase in the advertising budget). Unless we somehow take into account this interrelationship, we cannot isolate the 'true' impact of either marketing variable on sales revenue. For the same reason, we cannot compute the combined influence of advertising and personal selling on sales revenue from our bivariate results: it is not possible to simply add the r^2 values associated with advertising and sales people together because, if these variables are intercorrelated, each r^2 does not reflect *only* the relationship between the variable under consideration and sales revenue; part of each r^2 could be attributed to the other variable. Consequently, the proportion of variance in sales revenue that is jointly explained by advertising and personal selling is less than what would be obtained if the two r^2 values were simply added up (have a look at Figure 14.1 and you will see immediately what the problem is). In short, unless we use a procedure that investigates the relationship between sales revenue and the two marketing variables whilst considering the interrelationship among the latter (i.e. unless we analyse all three variables *simultaneously*), we are stuck.

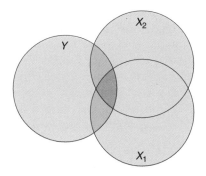

Figure 14.1 Relationships among three variables. Y = sales revenue; X_1 = advertising; X_2 = number of sales people.

It is for such reasons that we opt for multivariate techniques. Whenever we are facing a situation where multiple variables are involved and the potential interrelationships among these variables *must* be taken into account in order to be able to answer our research questions (i.e. variables are interrelated in such a way that their effects cannot be meaningfully interpreted separately), multivariate analysis procedures are our only option. Moreover, from a practical perspective, many problems are multivariate in nature (e.g. sales performance may be a function of several potentially interrelated variables; product choice may be determined by sets of interdependent attributes; market segments may differ in terms of several interlinked characteristics; and so on). Consequently, researchers who have an appreciation of multivariate techniques will obtain a more realistic understanding of complex problems.

> **HINT 14.1**
>
> Most complex (read: real life) problems are multivariate in nature. An understanding of multivariate analysis, although technically demanding, is likely to pay off in terms of better understanding of complex problems.

Types of Multivariate Techniques

Multivariate analysis has largely become popular through the widespread use of statistical computer packages (it is now probably as popular as Nintendo!). Mystical calculations that would literally have taken months to perform by hand (a task only willingly undertaken by the sort of people who leave parties early in order to have fun with a random numbers generator), can now be managed with a few simple computer commands producing near-instantaneous results. The techniques outlined below are available on practically all major commercial statistical packages (see Chapter 5 for details); specialized programs that focus only on one multivariate technique also exist, but you must be really in love with the technique (or a compulsive collector of statistical software) to justify buying them.

Multivariate techniques can be broadly divided into two main groups, namely **dependence methods** and **interdependence methods**. All

dependence methods
interdependence methods

dependent variables
independent variables

dependence methods are characterized by a distinction between **dependent variables** and **independent variables**. Dependent variables are those that are predicted or explained by (yes, you've guessed it) the independent variables. In our earlier example, sales revenue was the dependent variable and number of sales people and advertising were the independent variables. Interdependence methods do not make a distinction between dependent and independent variables; such techniques are frequently employed at an early stage of a data analysis, namely when it is still unclear which relationships and structures exist in a given data set (we are still debating whether the techniques listed in the *Kama Sutra* should be treated as dependence or interdependence techniques – any ideas?).

To identify which multivariate technique is appropriate in a given situation, it is important that you ask the right questions. First, you should ask whether your data permits you to distinguish between dependent and independent variables. If the answer is yes, you will need a dependence method (see Figure 14.2); if the answer is no, you will need an interdependence method (see Figure 14.3).

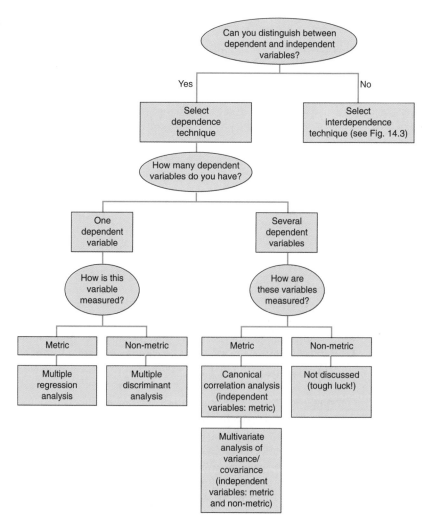

Figure 14.2   *Selecting multivariate techniques: dependence methods.*

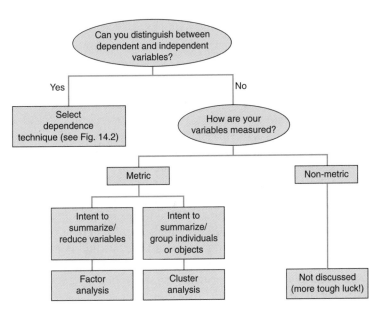

Figure 14.3 Selecting multivariate techniques: interdependence methods.

Focusing on dependence methods, the first issue you need to clarify is the *number* and *measurement level* of the dependent variables. If you are dealing with just one dependent variable and this one variable is measured on a metric scale, you most likely require a **multiple regression analysis**. If, in contrast, your one dependent variable is non-metric, you should have a look at **multiple discriminant analysis**. For both kinds of analysis, your independent variables should be metric (inclusive of dichotomous variables with dummy-variable scoring – see Chapter 3).

Turning to situations in which several metric dependent variables are involved, you will need either a **canonical correlation analysis** or a **multivariate analysis of variance**. For the former your independent variables must be metric, while for the latter your independent variables can be non-metric (usually nominal).

Compared to selecting an appropriate dependence technique (and compared to assembling some of the horrible flat-packed furniture you can buy these days), choosing a suitable interdependence technique is relatively easy. Essentially, you only need to answer one question: Do you want to summarize/reduce your *variables* into a smaller number of dimensions or do you want to summarize/organize your *units* (i.e. individuals or objects) into groups? Depending upon your answer to this question, you have a choice between **factor analysis** and **cluster analysis**.

In the next section, we provide a brief description of the techniques included in Figures 14.2 and 14.3; while these are the most useful and widely used techniques, there are many others available (see Further Reading). Note that, given the technical complexity of multivariate techniques, you should not expect to learn how to use them simply from the description that follows. While you can gain an initial understanding of the circumstances in which these techniques might be useful, there are no short cuts here: you *must* consult specialized textbooks (such as the ones recommended in the Further Reading section) if you want to become a master of multivariate analysis.

multiple regression analysis

multiple discriminant analysis

canonical correlation analysis

multivariate analysis of variance

factor analysis

cluster analysis

Dependence Methods

Multiple Regression Analysis

Multiple regression analysis is used to analyse the relationship between one dependent variable and a number of independent variables. Both the dependent and the independent variables need to be metric, i.e. measured at interval or ratio level (independent variables can also be in the form of 'dummies'). Assume that you would like to predict the amount of money people will donate to *Computer Addicts Anonymous* (a newly created charity) during the next year. Assume further that you expect three variables to influence the amount donated, namely the number of times people are asked to give money, the income of the potential donors and the number of computers in the family. Multiple regression analysis would be the ideal analysis technique for this task. It would not only enable the prediction of the dependent variable but also provide an assessment of the *relative* impact of each of the three independent variables; moreover, it would indicate the *combined* ability of the independent variables in explaining the variation in the dependent variable. Impressed? So are we!

Multiple Discriminant Analysis

Multiple discriminant analysis is used to analyse differences between groups in terms of several variables simultaneously. It is conceptually very similar to multiple regression analysis, the difference being that the dependent variable is now non-metric (i.e. a nominal variable defining group membership). For example, if you would like to find out whether donors and non-donors (a dichotomous variable) to *Computer Addicts Anonymous* differ in terms of age, number of children, number of computers owned, and number of computer magazine subscriptions, discriminant analysis would be the technique to use. The results would tell you whether donors and non-donors score differently on these variables, and also identify which variable is the best, second best, third best, etc., in terms of discriminating power. In addition, the results could be used for prediction purposes, i.e. to classify people, for whom you do *not* know whether they are donors or non-donors, into one of these categories based on knowledge of their income, number of children, computer ownership and number of computer magazine subscriptions. In most applications, the dependent variable in discriminant analysis is dichotomous (for example, buyers versus non-buyers, or hang-gliding instructors versus Sunday school teachers), but the technique is also applicable when a multichotomous dependent variable is involved (for example, heavy smokers versus casual smokers versus non-smokers).

Canonical Correlation Analysis

Canonical correlation analysis is a multivariate technique which can be used in situations with multiple dependent *and* multiple independent variables; both sets of variables have to be metric. This technique can also be viewed as an extension of multiple regression analysis, the key difference being the number of dependent variables. Canonical correlation analysis would, for instance, be useful if you wanted to simultaneously analyse the impact of several company characteristics (e.g. number of new product introductions during the last year, size of advertising budget, number of export personnel, and salary of chief executive) on several aspects of export performance (e.g. export sales, export profits, and export market share). What the technique would do is derive separate linear combinations of the independent and dependent variables in such a way so as to maximize the correlation between the sets of independent and dependent variables. Put differently, the technique would create a new set of composite variables (called 'canonical variates') and optimally correlate the dependent canonical variates to the independent canonical variates (it's really simpler than it sounds!).

Multivariate Analysis of Variance

Multivariate analysis of variance (or MANOVA for short) is one of the most versatile multivariate techniques and can be used in a variety of different situations, particularly in experimental designs. Here we simply look at MANOVA largely as an extension of the one-way ANOVA procedure we discussed in Chapter 12. In this context, MANOVA can be used to gauge the impact of several non-metric independent variables on two or more metric dependent variables. Assume, for example, that you have developed a new breath freshener for dogs (and it is high time that somebody developed such a product). Now, you would like to test the influence of a number of alternative package formulations (categorical variables) on both usage frequency and satisfaction level (metric variables). You have the choice between a blue package or a red package, a spray formulation or a liquid formulation, and two brand names: *Freshhound* and *Dogsmile*. Thus, there are $2 \times 2 \times 2 = 8$ possible combinations of packages (formed by 'crossing' the three independent variables) and you want to identify the optimal one on both usage frequency and satisfaction with the product. MANOVA would allow you to do this by indicating the relative impact of each packaging characteristic as well as potential interactions between pairs of characteristics (known as 'two-way interactions') and between all three characteristics at once (known as 'three-way interactions').

An extension of MANOVA is **multivariate analysis of covariance** (MANCOVA). MANCOVA is used when the confounding (i.e. unwanted) influence on the dependent variables of one (or more) uncontrollable independent variables (measured on a metric scale) needs to be removed from the experiment. If you think that the age of the dog, for example, could possibly influence the usage frequency of and/or the satisfaction level with the dog breath freshener, 'age of dog' could be treated as a 'covariate' and its effect removed before the effects of the categorical independent variables are analysed.

Note that you must really know your stats to set up a MANOVA or MANCOVA properly, as there are several conditions that have to be met in order to get the best out of the technique; this is not to discourage you from using it but to encourage you to use it *correctly*.

multivariate analysis of covariance

Interdependence Methods

Factor Analysis

Factor analysis refers to a range of techniques the aims of which are to describe a larger number of (metric) variables by means of a smaller set of composite variables (so-called 'factors') and to aid the substantive interpretation of the data. There are two main types of factor analysis with different objectives.

common factor analysis   **Common factor analysis** focuses on the *common* variance (i.e. the variance shared among the original variables) and seeks to identify underlying dimensions (known as 'common factors'). To the extent that subsets among original variables reflect a common 'core' (i.e. are measuring the same underlying construct), the derived dimensions should be meaningful and interpretable. The original variables can then be described in terms of the common underlying dimensions. Common factor analysis is particularly useful in the context of measure development, as it enables an assessment of the dimensionality of multi-item scales (for example, if a scale is truly unidimensional, all items comprising the scale should share a single common factor).

principal components   **Principal components analysis**, on the other hand, focuses on the
analysis   *total* variance (i.e. the entire variation in the data set) and seeks to reduce the original set of variables into a smaller set of composite variables (called 'principal components') which are uncorrelated to one another. Each principal component is formed by linearly combining the original variables, the overall objective being to explain as much of the original variance in the data as possible by few principal components. As an example of principal components analysis, consider 20 questions (variables) cast into a 5-point Likert scale describing attitudes towards pet iguanas. After running a principal components analysis, you may find that the computer program has reduced the original 20 variables into, for example, three components which together explain 70% of the variance in the *original* variables. By looking at the individual variables that are strongly associated with each component, you may find that the first captures attitudes towards 'companionship', the second attitudes towards 'iguana hygiene' and the third component attitudes towards 'iguana obedience'.

The key difference between principal components analysis and common factor analysis is that, in the former case, the sole aim is to reduce the original set of variables into a smaller set of composite variables (components); it is simply a data reduction technique and makes no assumptions regarding the underlying structure of the data. In contrast, common factor analysis focuses explicitly on the interrelationships among the original variables and seeks to describe them in terms of common underlying dimensions; thus, the focus is on explaining the patterns of relationships among the original variables by means of a factor structure. Both types of factor analysis are widely used, the choice between them being governed by the specific needs of the researcher (e.g. measure development versus data reduction).

Cluster Analysis

Cluster analysis is also a technique used to reduce complexity in a data set. However, while factor analysis is concerned with reducing the number of *variables*, cluster analysis seeks to reduce the number of *objects* (e.g. individuals, products, advertisements) for which measurements have been obtained. By looking at the similarities and differences between the scores observed on the variables of interest, each object is grouped with others having similar scores into what are known as 'clusters'. The latter are mutually exclusive groupings formed in such a way that objects within the same cluster are more like each other than they are with objects in other clusters. Cluster analysis is often used in market segmentation studies, where market segments are formed on the basis of a large number of customer characteristics, e.g. income, number of children, age, past purchase record, attitudes towards shopping, etc. By appropriately identifying 'similar' consumers on the basis of these variables, distinct groups (i.e. clusters) are formed which can then be targeted with appropriate marketing strategies.

Note that there are several different procedures available for deriving clusters and, therefore, the term 'cluster analysis' is used to refer to a multitude of methods.

Summary

This chapter has provided you with a brief insight into multivariate analysis. First, we looked at multivariate analysis as an extension of univariate/bivariate analysis and explained why multivariate techniques overcome the inherent limitations of simpler methods. Subsequently, we distinguished between two types of multivariate techniques, namely dependence and interdependence methods, and provided some guidelines for choosing between them. Finally, we gave non-technical descriptions of the most popular mutivariate procedures and the data analysis situations in which they may come in handy.

Questions and Problems

1. Explain, in general terms, the difference between univariate, bivariate, and multivariate analysis.

2. What can you do with multivariate methods that you cannot do with univariate and bivariate methods?

3. Explain the difference between so-called dependence methods and interdependence methods.

4. Which multivariate technique would you use to determine how two groups differ in terms of several variables?

5. Give an example of a research situation for which multiple regression analysis would be appropriate.

6. Which analysis is commonly used to correlate simultaneously multiple metric dependent variables with multiple metric independent variables?

7. What, if any, is the difference between common factor analysis and principal component analysis?

8. Which multivariate technique is widely used to classify people or objects into a smaller number of meaningful groups? Why would such a technique be useful?

9. When would you use multivariate analysis of variance (MANOVA)?

10. Haven't you had enough of data analysis yet?

Further Reading

Chatfield, C & Collins, AJ (1991) *Introduction to Multivariate Analysis*. London: Chapman and Hall. Slightly more technically complex than Hair et al. (1992). Introduces some of the mathematical concepts behind multivariate analysis.

Hair, JF, Anderson, RE, Tatham, RL & Black, WC (1992) *Multivariate Data Analysis*, 4th edn. New York: Macmillan. An excellent, non-technical introduction into multivariate analysis with good examples. Includes journal articles which illustrate the usage of the discussed techniques.

Sharma, S (1996) *Applied Multivariate Techniques*. New York: Wiley. A very well-written text of moderate technical difficulty. Best appreciated after reading Hair et al. (1992).

Chapter 15

It's all over . . . or is it?

Congratulations! You finally completed your path-breaking analysis of secret Internet usage in Amish communities (or whatever research project you have been working on). You squeezed every possible bit of information out of your data set. You compared your findings with your initial objectives, bearing in mind all the words of infinite wisdom contained in Chapter 6. You produced enough computer printouts to feel personally responsible for the deforestation of Latin America. Yes, you have truly *finished* your analysis! Great achievement! Go and celebrate! But not for too long, since an important piece of work is still ahead of you – the effective *communication* of your results.

Most researchers are, quite understandably, proud of their work. However, they are usually so closely involved with every intricate detail of their analysis that they sometimes find it difficult to keep in mind the 'big picture'. Consequently, they are at risk of confusing **statistical significance** with **theoretical importance** and/or **managerial importance**, find it hard to write punchy and relevant **research reports**, and have problems with making effective **oral presentations**. In what follows we give you some hints on how to cope with the above; after all what's the point of doing a fantastic analysis if nobody appreciates it?

statistical significance
theoretical importance
managerial importance
research reports
oral presentations

The Written Research Report

Before you write down the first line, you need to think about the requirements of your *audience*. If your research is part of your research thesis and you are presenting the findings to your supervisor, the sophistication and degree of detail required are likely to be relatively high. But while the technical sophistication of your audience determines the maximum depth of your report, you should not automatically assume that all readers who understand the technical details will also be *interested* in them or, indeed, have the time (or even the inclination) to read them. This is why you must make a distinction between the statistical characteristics of the results (e.g. whether they are significant or not) and their theoretical and/or managerial substance. Not every single result that is statistically significant is also important from either a theoretical or a practical point of view (see also Chapter 10, particularly Warning 10.7). Be very aware of this distinction and focus on those results that impact on management decision

WARNING 15.1

In reporting your findings focus on their theoretical importance and managerial relevance, not simply on their statistical significance.

making or, in an academic context, are important for theory development purposes. Also do not forget that it is often more important that a particular estimate or relationship is *not* significant. Thus, do not fall into the trap of reporting only findings that are significant; also discuss non-significant results that provide managerial or theoretical insights.

Whenever possible, you should try to find out which style and format your target audience prefers *before* you start writing up your results. Most companies like very short and succinct reports that focus on the managerial implications of the research findings. A PhD thesis, on the other hand, has to discuss methodological details in depth. Consequently, when you are in a situation where you have to please multiple audiences (for example, your thesis supervisor and a sponsoring company), it is best to produce separate reports for the different audiences you plan to address. Although this means extra work (and may involve getting up at the crack of dawn, i.e. at 10.30 a.m. instead of 11.00 a.m.), the effort usually results in documents that are far more useful for the different types of readers.

summary

Most reports should be preceded by a short **summary** (often referred to as 'executive summary'). In a corporate environment, this is often the only part which is read by all of your readers (people have *work* to do, you know!). Therefore, the summary may be the most important part of your research report and you need to ensure that it is 'polished'. A good summary has to be able to stand on its own, i.e. has to make sense without having to read the rest of the report (i.e. all 2300 pages of it).

introduction

Following the summary, your report should open with an **introduction**. Here, the main aim is to familiarize the reader with the purpose and scope of the research and place the study into context (provide some historical background, indicate whether similar studies have been conducted previously, discuss past findings, etc.). In marketing terms, the introduction is a *positioning* exercise in which the reader needs to be convinced of the need for and importance of your study.

literature review

In academic studies, the introduction is normally followed by a separate **literature review**, which extends the discussion of related research. If the pertinent literature is extensive, it is helpful to locate, and refer to, review articles (in such circumstances, it may also be useful to construct summary tables of the findings of previous studies).

The main body of the research report comprises details on methods, results and limitations. Again, there is a difference between academic reports and management reports. Academic reports almost always contain a separate **methodology** chapter. This would include a precise definition of the population that was studied, as well as detailed information on the sample size, sampling procedure, response rates, the research instrument (e.g. questionnaire or interview schedule) and the way the variables were measured. In a management report, information on the methodology is usually much less detailed, often included in the discussion of the results and frequently relegated to an appendix.

methodology
results

The **results** section (also often called 'analysis' or 'findings') is the place to stun your readers with all your nice tables, three-dimensional colour figures, and compelling arguments (at your discretion, you may also include poems or dried flowers!). Treat the results section as a *story* that needs to be unfolded in a *logical* sequence (i.e. do not first discuss the relationship between variable *X* and variable *Y* and *then* provide descriptive statistics on *X* and *Y*). Sometimes, you may want to use sub-headings to provide the reader with signposts indicating how your analysis has been structured. It is often suggested that the presentation of the results should be followed by a section on the study's **limitations**. This should draw the readers' attention to possible biases in the findings (for example the exclusive sampling of organic vegetable farmers in a survey aimed at identifying the most influential Rap group of the year) and clarify how far the findings can be generalized. It is usually more advantageous to be open and frank about the limitations inherent in a research study than leaving them to the reader to discover. In fact, an open admission of research limitations will demonstrate that you are honest and able to distance yourself from your work. Thus, it may actually increase the readers' opinion of the study, rather than lower it.

An alternative approach is to discuss the limitations of the study *either* in the methodology section *or* in the conclusion section. In the latter case, specific limitations can be linked to a discussion of **future research needs**. Remember, any investigation worth its salt always raises more questions than it answers (you do not want to lose your job as an analyst, do you?).

The **conclusions and recommendations** section rounds off the research report. A clear link should always be established between the original study objectives (as described in the introduction) and the respective conclusions. Sometimes, you may even (briefly) restate the original objectives of the study and then present the conclusions relating to each individual objective. In a commercial setting, studies are initiated to help make specific decisions and, therefore, your conclusions should demonstrate how the results of your analysis will help make those decisions. Similarly, in an academic environment, research is usually conducted to build theory. Here, it is crucial to show the implications of your findings for theory building.

Unfortunately, there is no one best way of structuring a research report; the nature of the audience, the research topic, space limitations, etc. will all have an impact on its final form. However, if only to illustrate different styles, Table 15.1 presents two 'generic' reports; the first is more suitable for an academic environment, while the other is more compatible with a company setting. These 'generic' reports are followed by a more 'action-oriented' report on a specific topic. Note that we have chosen relatively 'dry' headings for the first two examples. This is not to suggest that these types of reports cannot use more imaginative headings, but only to illustrate their structure more clearly.

> **HINT 15.3**
>
> The introduction section should state the aims and the scope of the research and position it against similar/previous studies.

limitations

future research needs

conclusions and recommendations

> **HINT 15.4**
>
> Always demonstrate how the results of your analysis impact on the decision(s) management needs to make or, in an academic setting, contribute to theory building. Make sure that what you 'promise' in the introduction section tallies with what you 'deliver' in the conclusions section!

The Oral Presentation

In addition to one or more written reports, you may also have to give an oral report of your research. In fact, oral presentations are quite common at the **research proposal** stage and in the form of **progress reports** at various stages during lengthy projects. Depending on the specific corporate environment or academic setting, the formal oral report at the conclusion

research proposals
progress reports

Table 15.1 Examples of research reports

Generic academic report

1. Title page
2. Table of contents
3. Summary
4. Introduction
5. Literature review
6. Methodology
7. Findings
8. Conclusions and recommendations
9. Limitations and future research
10. Bibliography
11. Appendices

Generic company report

1. Title page
2. Table of contents
3. Executive summary
4. Introduction
5. Body (results)
6. Conclusions and recommendations
7. Appendices (including methodology)

Action-oriented report

1. The Amish Internet secrets (title)
2. The shocking facts (summary)
3. Why the world needs to know (introduction – positioning)
4. How we revealed the truth (methodology)
5. Inside the secret Amish computer camp (body)
6. What the Amish download (body)
7. Amish software products (body)
8. An untapped market opportunity (conclusion)
9. How can we reach the New-Age Amish (recommendations)

WARNING 15.2

Never underestimate the importance of an oral research report.

of your project may precede or follow the distribution of the written report(s). In any event, it is important to recognize that many senior managers (and even some busy academics) will judge the quality and usefulness of your work virtually exclusively on the basis of your oral presentation – perhaps supported by a quick glance at the summary and/or conclusions section of your written report over a cup of coffee. While this insight can be quite disheartening (particularly if you have been working full-steam for five years to complete your study), it follows that oral reports are of utmost importance and it is in your interest that they are done right.

So what should you do to make your oral presentation as effective as possible? The first step is no different from preparing your written report, namely that you get as much advance information about your likely audience as possible. Find out about your audience's interest, likely involvement and technical (statistical) sophistication level. Find out about the likely setting. Will your presentation be held in a small conference room or in a large auditorium? Ask yourself which parts of your research are likely to be the most interesting, relevant and important *from the audience's point of view*.

Next, you should develop a clear objective for your presentation. Usually, you want to persuade an audience that your findings do or do not support certain decisions/theories. Consequently, you need to carefully select the **key arguments** that best support your conclusions. The emphasis here is on *selection*; do not try to present everything you have done. A far more effective approach is to focus on no more than three key arguments (people forget and get bored quickly). For each key argument, you should subsequently consider which evidence (finding) is most suitable to support it. Try to illustrate the point with audiovisual material, draw comparisons, give examples, or use authoritative quotes; above all, *keep your audience interested*.

In contrast to written reports, there are no obvious headings and sub-headings in an oral presentation. It is therefore extremely important that you set clear **markers** to ensure that the audience notices when you have finished one point and are moving on to the next. Usually, these are short but important 'linking' sentences such as 'Having discussed the organization of the secret Amish computer camp, I shall now tell you which software they are using.' All too often, markers are overlooked by the presenter. This results in the audience getting lost and wondering what a particular point has to do with what you said before.

With your objectives firmly established and the key points supporting your objectives developed, your presentation can now take shape. It is always helpful to remember the old adage: 'Tell them what you are going to tell them (introduction), tell them (core presentation), and tell them what you told them (summary).' Thus, you can think about the introduction as providing a preview and the summary as providing a review. The only other two points to focus on now are (a) finding an attention-grabbing opener to your presentation (for example, a quote or an anecdote) and (b) ensuring that the audience clearly understands what actions flow from your report. Table 15.2 shows a suggested structure for an oral presentation.

Having finalized your oral report, it is imperative that you practise it a few times. Ideally, you should learn as much as possible of your oral report off by heart so that you can *talk to your audience*; at a minimum, you should know the beginning and the end. Under no circumstances should you *read* your report (after all, your audience can probably read just as well as you can!). If you must, have a few notes on index cards for consultation but refrain from using a full-text copy of your presentation.

Before leaving the discussion of oral presentation issues, we should emphasize that *visual aids* are absolutely essential for effective communication. Given today's technology, there is really no longer any excuse for handwritten overhead slides or for the use of a chalkboard. Instead, you should use coloured overhead transparencies or a computer-aided presentation prepared with the help of graphics packages (see Chapter 5).

A good oral presentation is as much of an art as it is a science. In the framework of this book, we can obviously only scratch the surface of the topic. If you think you need more help, we strongly advise you to buy one

HINT 15.5

In preparing an oral presentation, put yourself in the shoes of your audience. Keep the needs of your audience in mind and you will not go wrong.

key arguments

markers

WARNING 15.3

Do not forget to use appropriate 'markers' throughout your presentation. Otherwise, you risk losing your audience.

WARNING 15.4

Do not read your report when making an oral presentation – it is a sure way of boring/alienating your audience.

Table 15.2 Suggested structure for an oral presentation

HINT 15.6

Nicely prepared audiovisual aids will always help and never harm an oral presentation.

1. Introduction
    (a) Attention-grabbing opener
    (b) State objectives of your presentation
    (c) Tell them what you are going to tell them

2. Core presentation (tell them)
    (a) Supporting evidence no. 1
    (b) Marker to clarify the structure
    (c) Supporting evidence no. 2
    (d) Marker to clarify the structure
    (e) Supporting evidence no. 3
    •
    •
    •

3. Summary
    (a) Review: tell them what you told them
    (b) Indicated actions: tell them what they need to do

of the numerous books on presentation skills (these also usually include guidelines on the preparation of visual aids, voice projection, eye contact, etc.). You can also join public speaking classes at your local adult education college or join Toastmasters, a group that meets specifically to develop the public speaking skills of its members (giving impromptu talks on your research at your local supermarket or sauna club is also an alternative, albeit less conventional!).

Well, this is it! Nothing more. No more hints. No more warnings. No more statistics. And no more jokes! We profusely apologize to all readers and book reviewers who found our jokes immature, politically incorrect, or downright silly. We also apologize for having a go at statisticians (we just couldn't resist it!). Finally, we apologize to our editor for all the hassle we have given her (sorry Jennifer, we *do* love you).

You are now on your own. Hopefully, we have convinced you that you do not need a PhD in mathematical statistics to understand and conduct sensible data analysis. Perhaps we have even managed to demonstrate that learning about data analysis can be fun. But, above all, we hope that we have been successful in **Taking the fear out of data analysis**. Take care.

Summary

This final chapter demonstrated that successful data analysis should not end with a pile of computer printouts. Only a professionally produced research report and a punchy and eloquent oral presentation will ensure that your data analysis efforts are fully appreciated. The key to successful reports, both written and oral, is an understanding of the audience's perspective. This requires a knowledge of its technical sophistication, interests and priorities. We have given you some suggestions on how to structure written reports and some hints on how to make effective oral presentations. By now, you should be able not only to conduct a brilliant piece of data analysis but also to convince other people that you have done a wonderful job!

Questions and Problems

1. Provide an example illustrating the difference between statistical significance and theoretical or managerial substance.

2. Why does it sometimes make sense to report relationships which are *not* significant?

3. How does the nature of the audience affect the content and structure of a written research report?

4. Describe, in general terms, how you would structure (a) an academic report, and (b) a managerial report.

5. In your opinion, which is the most important part of a written research report? Why?

6. What advice would you give to someone preparing for his/her first oral presentation?

7. Why shouldn't you just read your written report to your audience?

8. Why are 'markers' important in oral presentations?

9. Draw up a list of everything you think can go wrong with an oral presentation.

10. Did you enjoy this book?

Further Reading

Arredondo, Lani (1991) *How to Present Like a Pro: Getting People to See Things your Way*. New York: McGraw-Hill. Very 'hands-on' pocket book with an emphasis on oral presentations.

Gallagher, WJ (1969) *Report Writing for Management*. Reading, MA: Addison-Wesley. A classic on the subject – well worth a look despite its age.

Morris, D & Chandra, S (1992) *Guidelines for Writing a Research Report*. Chicago: American Marketing Association. A good source of advice on how to structure and present research reports from a marketing perspective.

Index

Note: Page numbers in **bold type** indicate the first (or main) mention of a keyword or term.

absolute frequencies **74**
absolute magnitude 26
absolute zero point **26**
abstracting & index services **5**
acceptance region **143**
age measurement 27–8
alphanumeric variable **43**
alternative hypotheses **132, 145**
   formulation 136–7
ambiguous answers **40**
analysis **62**
   choice of method 66–71
   content of **64**
   independent v. related measures 69
   non-parametric methods 70
   objectives 62–3
   parametric methods 70
   setting objectives 63–4
annual reports 5
arbitrary zero point **26**
arithmetic average **90**
ASCII **56**
associations 134
   see also measures of association; relationships
asymmetrical distribution 91
attitudes **4**
attitude-measurement techniques 29
averages **90**
   measures of 100
awareness **4**

back-up copies 60
balanced v. unbalanced distribution of response
     alternatives 32
bar chart **82**
Bayesian approach **17**
BestFit 56
best-fitting normal distribution **157**
between-groups mean square **189**
between-groups sum of squares **189**
biased estimate **103**
bimodal distribution **94**
binomial distribution 107, 123

binomial test **166**
bivariate analysis **67**
bivariate data analysis **3**
bivariate normal distribution 205
bivariate relationships **198**

canonical correlation analysis **213**, 214–15
category nominal scale **24**
causality 206–7
cell **54**
census **10**
central location **90**
   measurement 93–100
central office edit **40**
Chebychev's theorem **106**
chi-square distribution **107**, 165
chi-square statistic 200
chi-square test 146
class interval **77–8**
class limits **77–8**
class midpoint **78–9**
class width **78**, 80
classes **77–8**
cluster analysis **213**, 217
Cochran Q test **193**
code book **43**, 47
coding forms **48**
coding template **48**
coefficient of kurtosis **105**
coefficient of skewness **105**
coefficient of variation **104**
collectively exhaustive categories 24, 78
column number **46**
common factor analysis **216**
comparability of responses **69**
comparisons 172–97
   between groups 173–90
   of variables 190–5
   statistical techniques for 174
competing hypothesis **132**
composite scale **105**
computer packages 126, 204
   types and applications **53–4**

computers in data analysis 56–8
concept **21**
conceptual definition **21–2**
conclusions and recommendations 221
confidence **17**
confidence coefficient **120**, 124
confidence intervals **65**, 119, 122, 126, 147–8
    setting 119–21
confidence level **119**
confidence limits **119**
constant 2
construct **21**
consumer reaction measurement 42
content of analysis **64**
contingency table **175**, 180
continuous frequency distribution **84**
continuous probability distributions **108**
continuous variable **77**
correlation 206–7
correlation matrices **204–5**
Cramer's *V* **199–201**
critical value **120, 143**
cross-classification **17**
cross-sectional data **5**
cumulative frequency 154
cumulative frequency distributions **74**, 85
    examples 76
cumulative frequency polygon 85

data 1
    and information 7–8
    types 4–7
data analysis
    computers in **54–5**
    software selection 59
data analysis packages 56–8
data cleaning **39**
data coding **43**
    mistakes **48–9**
data description, purposes of 73–4
data entry 48
    mistakes 48–9
data entry package **55**
data input package 54–6
data matrix **2**, 21
    computer-readable 43
    general form 3
    variable location and length in 46
data sets 2, 6
database services 5
degree of dispersion 152
degree of variability 164
degrees of freedom **108**
dependence methods **211**, 214–15
dependent variables **212**
descriptive analysis, purpose of 73–4
descriptive focus **64**
descriptive statistics **65**
desktop publishing 58
dichotomous variables **30**

directional hypotheses **133**, 144
Directory of Software for Marketing & Marketing
    Research 59
discrete probability distributions **108**
discrete variable 77
distribution of values in the population 69–70
dummy-variable coding **30**

editing **40**
effective sample size **69**
empirical distributions **108**
equality of intervals **25**
equivalence **24, 35**
errors in hypothesis-testing 138
estimation 116–29
    focus **65**
    nature of 116–19
    steps involved 127
Excel **55**
expected cost of errors **14**
experimental design **7**
experiments 5
exploratory hypotheses **133**, 145

F-distribution **107, 185**, 189
F-ratio 189
F-test for equality of variances **185**
factor analysis **213**, 216
facts **4**
field edit **40**
finite population **12**
    correction **123**
first quartile **76**
Fisher's exact test **178**
fit assessment 153–9
fixed sample **18**
focus of analysis **64**
Forecast Pro 56
Freelance 58
frequency distribution **74**
    characterizing 90–3
    graphical representation 81–8
    grouped 77–81, **78**
frequency polygon **83–5**
Friedman two-way analysis of variance (ANOVA) **194**
future research needs 221

generalizability **36**
goodness-of-fit tests **152**, 154
graph construction guidelines 87
graphical representation **81**
group comparisons 172, 173–90
grouped frequency distribution 77–81, **78**
    guidelines 79

H statistic 184
Harvard Graphics 58
histogram **83–5**
homogeneous population **14**

hypergeometric distribution **107**, 178
hypotheses
    single sample 152–3
    types 135
hypothesis-testing **130–5**
    and confidence intervals 147–8
    error in 138
    focus **65**
    general approach 135–47
    steps involved **135–47**

independent measures **68**, 172, 190
independent v. related measures analysis 69
independent variables 212
index of diversity **101**
inferential statistics **65**
information **7**
integrated packages **60**
intentions **4**
interdependence methods **211**, 216–17
Interleaf 58
interquartile range **102**
interval estimate **117**
interval scale **25**
interviewer number **40**
intransitive responses **41**
introduction 220
item non-response **41**
itemized rating scale 29

k-sample chi-square test 179
Kendall's rank-order correlation 202
Kendall's tau-c 202
key arguments 223
knowledge **4**
Kruskal-Wallis one-way analysis of variance **182**
kurtosis 92

label nominal scale **24**
laws of chance **13**
least squares **98**
leptokurtic distribution **92**
level of measurement **3, 24**, 28, 67, 141
level of significance 145–6
likelihood ratio 177
Likert scale **29**, 87
limitations 221
linear relationships **203**
literature review **220**
logically inconsistent data **41**
longitudinal data **5**
Lotus 1-2-3 55

McNemar test **192**
managerial importance **219**
Mann–Whitney *U* test 180, 191
Mantel-Haenszel test 177
markers **223**
mean **26**, 97, 99
measurement, nature of 21–30

measurement characteristics 67
measurement error **32**
measurement quality **32**
measurement rules **22**
measurement scales 23
measures of association **198**, 199–206
median 25, 95–7, 99, 159–60
mesokurtic distribution **92**
methodology **220**
metric data **27**
minimum/maximum values 189
MINITAB 57
missing data **41**, 47
missing values **41**, 47
mistakes 48–9
mode 25, 93, 99
motives **5**
multimodal distributions **94**
multiple comparison tests **190**
multiple discriminant analysis **213**, 214
multiple measures 172
multiple regression analysis **213**, 214
multiple responses **3**
multivariate analysis **67**, 209
    nature of 209–11
multivariate analysis of covariance (MANCOVA) **215**
multivariate analysis of variance (MANOVA) **209**,
      213, 215
multivariate data analysis **3**
multivariate hypothesis **210**
multivariate relationships **198**
multivariate significance tests **210**
multivariate techniques, types 211–13
mutually exclusive categories 24, **78**

negatively skewed distribution **91**, 99
nested measurement scales 27
nominal scale **24**
nonlinear relationships **203**
non-metric data **27**
non-overlapping fashion 78
non-parametric methods of analysis 70
non-parametric statistical techniques **27, 67**, 69
non-parametric tests **141**
non-probability sampling **13**
non-resistant measure **97**
non-response **17**
non-sampling errors **13**
normal curve **107**
normal distribution 70, 92, 107, 156
null hypothesis **132**, 144, 145, 146, 164, 169,
      181, 182, 191
    formulation 136–7
    rejection or non-rejection 146–7
null results **149**
number of sub-samples 67
number of variables 67
numeric variable **43**

observation methods **5**

observed frequencies **154**
observed score 32
odd v. even number of response alternatives 32
ogive **86**
omnibus panel **7**
one-sample chi-square test **154**
one-sample Kolmogorov-Smirnov test **154**, 157–9
one-sample proportion tests 166
one-sample sign test **159**
one-sample *t*-test **159**, 161–4
one-sample *z*-test 163
one-tailed test **145**, 146
one-way analysis of variance (ANOVA) 187–90
open-ended interval **81**
operational definition **22**
opinions **4**
optical scanner **48**
OR/MS Resource Directory 59
oral presentations **219**, 221–4
ordinal scale **25**
out-of-range values 74
outliers **97**

p-value **146**
Pagemaker 58
paired-sample sign test **191**, 193
paired-sample *t*-test 191, **195**
panel data **7**
parametric methods of analysis 70
parametric statistics 27, **67**, 69
parametric tests **141**
partitioning 189
Pearson's product moment correlation **199**, 203–6
percentiles **76**
phi coefficient **200**
pie chart **82**
platykurtic distribution **92**
point estimate **116**
Poisson distribution **107**
population 2, 10
population characteristics 153
population elements **12**
population estimates **14**
population mean, estimation 123–6
population parameters **92**, 116, 119, 148
  estimation 126–7
population proportion, estimation 121–3
positively skewed distribution 91, 99
power of a test **142**
Power Point 58
precision **17**
presentation packages 58–60
primary data **5**
principal components analysis **216**
probability distributions **108**, 114
probability sampling 13, 14
process of measurement **22**
progress reports **221**
proportion of variance **204**

published statistics **5**

qualitative variables **27**
quantitative variables **27**
quartile **76**
Quattro Pro **55**
questions 3, 4

random error **33**
randomness 167–9
range **101**
ratio scale **26**
raw data **8**
rejection region **143**
related measurements **68**, 172, 190–5
relationships 134, 198–208
  among variables 211
relative frequencies **74**
reliability **33**
  assessment 35
representative sample **13**
research proposals 221
research purpose **63**
research reports **219**
respondent number **40**
results 39, **220**
robustness test **142**
runs **168**

sample 2, 10
sample elements **12**
sample independence **68**
sample selection **12**
sample size **12**, 66, 124, 141
  determination 16–18
sample statistics **92**, 116, 121
sampling, nature of 10–12
sampling distribution of a proportion **118**
sampling distribution of the mean **124**
sampling distributions, theoretical **118**
sampling error 12, 117
sampling frame **14**
sampling methods 14–16
sampling procedure 13
sampling process 19
sampling units **16**
SAS 57
scaling formats **30–2**
scattergram **203**
Scheffe test **190**
second quartile **76**
secondary data **5**
semantic differential scale **29**
sensitivity **36**
sequential sample **18**
significance level **139**
  specification 137–40
significance testing **139**
significance tests 65, 139
significant region **144**

simple random sample **127**
single sample hypotheses **152–3**
single-variable hypotheses 135
skewness **91**
small-sample statistics 67
software selection 59
Spearman's rank-order correlation **199**, 201–2
Spearman's rho 201–2
spreadsheets **54**
SPSS Data Entry **55–6**
squared deviations **98**
stability **35**
standard deviation **103**, 109–14, 118, 189
standard error 121, 127, 189
standard error of the mean **124**
standard error of the proportion **118**
standard normal distribution **108**, 110, 112, 114
standard scores **104**
standardization **104**
Stapel scale 29
stated class limits **78**
Statgraphics 57
statistical inference **65**
statistical significance 148–50, **219**
statistical tables **109**, 122
statistical tests 139
    selection 140–2
Student-Newman-Keuls test 190
Sturges' rule 80
substantive significance **150**
summary **220**
summary measures **90**
Supertree 56
surveys 5
symmetrical distribution **91**, 99
syndicated services **5**
system missing value 42
systematic error **33**

t-distribution **107**, 124–5
t-values **125**
temperature scales 25–6
test statistic **143**
    computation 146–7
test variables **48**
tests for location **152**, 159–64
tests for proportions **153**, 165–7
tests for randomness **153**, 167–9
tests for variability **152**, 164–5
tests of homogeneity **180**
tests of independence **180**
text editor **56**
theoretical distributions **108**
theoretical frequencies **154**
theoretical importance **219**
theoretical sampling distributions **118**

top quartile **76**
t_r-test for all pairs **196**
trend data **6**
true class limits **78**
true score **32**
Tukey-b test 190
two-sample chi-square test **175**
two-sample t-test **184**, 191
two-tailed test **145**
Type I error **138**
Type II error **138**

unbiased estimate **103**
unequal class intervals **80**
uniform distribution **93**
unimodal distribution **95**
units of analysis **1**
univariate analysis **67**
univariate data analysis **3**

validity **33**
    assessment 34–6
value labels **45**
value of information **17**
values **1**
variability **16**, **90**
    measuring 100–5
variable labels **45**
variable name **43**
variable transformations **49**
variables **1**, 3, 4
    recording 50
    relationships among 211
    types **46**
variance **102**
variance test **164**
Ventura 58
verification 149

Wald-Wolfowitz runs test **167**
Wilcoxon rank sum W test 180
within-group mean square **189**
within-group sum of squares **189**
Word 58
word processing packages 58
word processor **56**
Wordperfect 58

Yates' correction for continuity **177**

z-scores 123, 124, 143
z-statistic **143**
z-test **142**, 143, 146
z-test for a proportion **166**
z-test for differences in means **186**
z-test for differences in proportions **179**
zero point 26